Handbook of Repair and Maintenance of Domestic Electronics Appliances

As per revised syllabus of Modular Employable Skills Scheme

Course Code- ELC-701

AF623392

Er. Shashi Bhushan Sinha
Chief Assessor, Technomed Services

BPB PUBLICATIONS

First Edition 2016
REPRINT 2025

ISBN: 978-81-8333-502-7

Distributors:

COMPUTER BOOK CENTRE
12, Shrungar Shopping Centre,
M.G.Road, BENGALURU–560001
Ph: 25587923/25584641

MICRO BOOKS
Shanti Niketan Building,
8, Camac Street, KOLKATA-700017
Ph: 22826518/22826519

DECCAN AGENCIES
4-3-329, Bank Street,
Hyderabad-500195
Ph: 24756967/24756400

MICRO MEDIA
Shop No. 5, Mahendra Chambers,
150 DN Rd. Next to Capital Cinema,
V.T. (C.S.T.) Station, MUMBAI-400 001
Ph: 22078296/22078297

BPB BOOK CENTRE
376 Old Lajpat Rai Market,
Delhi-110006
Ph: 23861747

INFOTECH
G-2, Sidhartha Building,
96, Nehru Place,
New Delhi -110019
Ph: 41619735, 26438245

BPB PUBLICATIONS
20, Ansari Road, Darya Ganj
New Delhi-110002
Ph: 23254990/23254991

Published by Manish Jain for BPB Publications, 20, Ansari Road, Darya Ganj, New Delhi- 110002 and Printed at Manipal Technologies Limited, Manipal

Table of Contents

Preface

Technomed Services is empanelled as Assessing Body for Assessment of Trained candidates in the trades of Electronics, Electrical, Fabrication, Information & Communication Technology, Refrigeration & Airconditioning, Medical & Nursing and Allied Healthcare.

After the revision of the syllabus in 2014 a need was felt for developing guidebooks covering the study materials of MES Courses as per the new syllabus. It was also felt that there is a need to develop e-book so that it is easy to use the book in Electronic form to study anywhere and is accessible quickly and on demand.

It is with these objective that this " Handbook of Repair & Maintenance of Domestic Electronics Appliances for ELC-701" has been developed. The book first the series of e-books that will be developed by Technomed Services for various Modular Employable Skills courses.

The book has already been submitted for Registration of Copyrights under Copyrights Act to the Registrar of Copy rights in India.

The book has been written simple English and will be useful for the Trainees, Trainers as well as Assessors.

All our assessors are requested to keep a copy of this book for Assessment.

Objectives of the Book

This book covers the entire syllabus of ELC-701- Repair & Maintenance of Domestic Electronics Appliances. After getting the skills the trained candidates will be able to do the following:

- Testing of electrical parameters, cables and measurements.
- Select the proper instrument for suitable measurement.
- Identify and test passive and active electronics components.
- Practice soldering and de-soldering of various types of electrical and electronics components.
- Trouble shoot the faults in the given power supply circuits.
- Identify various Input and output sockets/connectors of the given UPS.
- Identify various functional blocks/major components/ICs in the given UPS.
- Monitor, measure major test points and Test the capacity of the given UPS and rectify the faults.
- Identify and test various mechanical and electrical modules of the given appliances.
- Identify electronics parts/components/modules of the given appliances.
- Aware of models of different appliances and features.
- Use reference manuals and identify the information required to service the appliances.
- Practice the standard troubleshooting procedures as suggested in the product manuals.
- Install the given appliance as recommended by manufacturer.

General information for Domestic Electronic Appliances

Name of Sector: ELECTRONICS

Name of Module: Repair and maintenance of Domestic Electronic Appliances

MES Code: ELC701

Duration of Course: 520 Hrs

Entry Qualification of Trainee: 8TH Pass + 14 yrs of age

Unit size (No. Of trainees): 20

Power Norms: 2.0 KW

Space Norms: (Workshop and 60 Sq.mtr Minimum size of one side Class Room) to be 04m.

Instructors Qualification: B.E./ B.Tech. in Electronics /Electronics& Telecommunication/Electronics& Communication with one year experience in the relevant field

OR

Diploma in Electronics/Electronics & Telecommunication / Electronics & Communication from recognized board of technical education with two years experience in the relevant field.

OR

NTC/NAC in the relevant trade with three years' experience respectively in the relevant field Desirable Craft Instructor Certificate (CIC)

Course Contents for Module: Repair and Maintenance of Domestic Electronic Appliances

Practical Competencies	Theory
Basics of electricity and Electrical Cables • Practice procedure for electrical and personal safety measures • Identify the Live, Neutral and Earth on power Socket • Construct a test lamp and use it to mains healthiness • Use a Tester to monitor AC power. • Measure the voltage between the neutral and ground and rectify earthing. • Identify and test different ac mains cables • Skin the electrical wires /cables • Measure the gauge of the wire using SWG • Make the mains cable for termination • Identify the primary and secondary cells • Measure and test the voltages of the given cells/battery using analog and digital Multi-meter • Charge and discharge the battery • Use a hydro-meter to measure the specific gravity of the secondary battery	Basic terms electric charges, Potential difference, Voltage, Current, and Resistance. Basics of AC & DC. Terms such as positive cycle, negative cycle, Frequency, Time period, RMS, Peak value , peak to peak, instantaneous values, Insulators, conductors and semiconductor. Different type of electrical cables and their specification.
Multi-meter • Cleaning the switch contacts using switch cleaning solution. • Testing the fuse. • Use the analog and digital Multi-meter to measure the DC voltage by doing measurement at the test points provided.	Multimeter- Precaution to be taken in handling an analog Multi-meter. Study of different controls on Multimeter Principle of operation of Analog Multi-meter. Precaution to be taken in handling digital Multi-meter Frequently occurring Problems in Analog multimeters and the remedial measures

Contd...

• Use the analog and digital Multi-meter to measure AC voltage measurement by doing measurement at the test points provided. • Adjust the zero adjustment screw for proper zero setting with the help of a screw driver before using Multi-meter. • Replace the battery in the Multi-meter. • Replacing the open Fuse with correct rating	Precaution to be taken in handling digital Multi-meter, Familiarization with operation controls of digital Multi-meter Principle of operation of digital Multi-meter. Frequently occurring problems in Digital multimeters and the remedial measures.
Basic Electronics Components • Identify the different types of resistors • Measure the resistor values using colour code and verify the reading by measuring in multi-meter • Identify the power rating using size • Identify different inductors and measure the values using LCR meter • Identify the different capacitors and measure capacitance of various capacitors using LCR meter • Dismantle and identify the different parts of a relay • Identify different types of mains transformers and test them • Identify the primary and secondary transformer windings • Measure the primary and secondary voltage of different transformers	**Basic Electronics Components** Resistor-definition, types of resistors, their construction & specific use, color-coding, power rating. Types of inductors, specifications and applications. Types of capacitors, specifications and applications Working principle of a Transformer, Specifications of a transformer, Step-up, Step down and isolation transformers Fuse – types, use of fuses and its rating.
Power supply • Testing of active components • Practice soldering and de-soldering techniques • Assemble and test– half wave, full,wave & bridge rectifier circuits with and without filter. • Identify the different types of fixed positive and negative regulator ICs (78/79 series) • Identify the pins • Construct a fixed voltage regulator using 78xx/79xx series ICs	Basic Electronics active components, testing of components, Working of half wave, full wave and bridge rectifier circuits. Voltage Regulator circuit. Applications of transistor – its uses MOSFET – precautions when handling. DIAC, SCR, TRIAC - application Identify the pin diagram of Voltage regulator ICs Applications of transistor – ts uses MOSFET – precautions when handling. DIAC, SCR, TRIAC - application identify the pin diagram of Voltage regulator ICs

Contd...

• Construct a variable voltage regulator using LM 723. • Observe the output voltage of different IC regulators by varying the input voltage	
UPS/Inverter • Installation of UPS and Inverters • Maintenance of batteries • Dismantle the UPS and identify the major parts • Testing of major components • Testing of power modules • Charging, discharging and testing of batteries.	Various types of batteries used in UPS and Inverters and their maintenance. Different types of inverter, UPS, Working principle, specifications, explanation with the help of block diagram.
SMPS • Dismantle the given SMPS and find major sections/ ICs components. • Measure voltages at vital points • Identify various input and output sockets/ connectors of the given SMPS. • Repairing of SMPS, simulating various faults diagnosing and rectifying it.	Block Diagram of Switch mode power supplies and their working principles.
Washing machine • Installation of front load washing machine • Installation of top load washing machine • Identify the internal and external parts of semi-auto washing machine • Identify the internal and external parts of fully automatic washing machine • Operate semi–automatic washing machine • Operate fully–automatic washing machine • Rectify the fault leading to not working of control panel switches. • Rectify the fault leading to not working of pulsator / agitator. • Rectify the fault leading to spin drier not working. • Rectify the fault leading to one side rotation of motor. • Rectify the fault leading to water inlet and outlet valves.	Washing M/c: different types of machines, washing techniques, parts of manual, semi automatic and fully automatic machines, basic working principle of manual, semi automatic and fully automatic machines, study the working of motors, different types of timers, power supply circuits.

Contd...

<table>
<tr><td>Microwave oven
• Identify the internal and external parts of micro wave oven.
• Identify the different touch pad controls their functions
• Testing of high voltage diode.
• Identify the HV capacitor and discharge it.
• Rectify the fault leading to fuse blows off when cooking is initiated.
• Rectify the fault leading to not responding of touch switches. (front panel)
• Rectify the fault leading to dead set.
• Rectify the fault leading to long cooking time.
• Precautions–importance of interlocking switch in performing maintenance</td><td>Microwave oven: Different types of oven, study the various functions of Oven, Electrical wiring diagram of microwave oven, working of Power supply.</td></tr>
<tr><td>Electric Iron
• Steam Iron
• Dismantle and identification of various
• parts, wiring, tracing of various controls,
• Electronic circuits in steam Iron
• Identify the faults in steam iron & rectify</td><td>Principle of electric iron, parts of steam iron, thermostat heat controls.</td></tr>
<tr><td>Electric Rice cooker
• Identify various components of Electric
• rice cooker, controls and trace the circuit
• and rectify the simulated faults.</td><td>Principle of working of rice cooker. Various parts & functions of rice cooker, temperature control and timer unit.</td></tr>
<tr><td>Electric kettle
Identify various components of Electric kettle , controls and trace the circuit and rectify the simulated faults.</td><td>Principle of working of electric kettle. Various parts & functions of Electric kettle and temperature control unit.</td></tr>
<tr><td>Mixer & Grinder
• Dismantle and identification of various parts, wiring, tracing of various controls,
• Electronic circuits in various types of Mixers/grinders
• Identify the faults in various types of Mixers/grinders & rectify</td><td>Various parts & functions of Mixer/ Grinder, speed control circuit & automatic over load protector.</td></tr>
</table>

CHAPTER 1

Basic Terms

BASICS OF ELECTRICAL SAFETY AND PRACTICAL PROCEDURES FOR ELECTRICAL AND PERSONAL SAFETY MEASURES

Electric charge is created when the Electrons accumulate at any device. A positive charge is created when there is a deficiency of Electrons and when there is an excess of electrons a negative charge is created.

The negative charge creation makes Negative potential whereas positive charge creation makes Positive potential.

The flow of charge is known as Electricity and it flows form Positive potential to Negative potential.

Please note that the electrons flow from negative potential to positive potentiaL but the direction of flow of electricity is from positive to negative potential. The conventional electricity flows opposite to the Flow of Electrons.

The Voltage is the difference of Charge and is also known as Potential difference.

The Current is the rate at which the Charge flows.

The Resistance is the opposition to the Flow of charge by an element in the path of flow which is also known as Circuit.

An analogy is made with Flow of water from a height. As water flows from a greater height to a lower height so the Charge flows from Higher potential to lower potential.

UNITS OF CHARGE, VOLTAGE, CURRENT AND RESISTANCE

Charge is measured in Coulombs

Voltage is measured in Volts

Resistance is measured in Ohms

Current is measured in Amperes

Ohms Law- Is the basic Law of Electricity flow and is as below:

$$V=I \times R$$

Where V is Voltage in Volts, I is Current in Amperes and R is Resistance in Ohms. Thus a Current of 1 Ampere flowing through a device of 1 Ohms will generate a Voltage of 1 V.

Basics of AC & DC. Terms such as positive cycle, negative cycle, Frequency, Time period, RMS, Peak value , peak to peak, instantaneous values,

Electricity flows in two ways: either in an **alternating current (AC)** or in a **direct current (DC)**. As already discussed the Electricity or "current" is nothing but the movement of electrons through a conductor, like a wire. The difference between AC and DC lies in the direction in which the electrons flow. In DC, the electrons flow steadily in a single direction, or "forward." In AC, electrons keep switching directions, sometimes going "forward" and then going "backward."

Alternating current is the best way to transmit electricity over large distances.

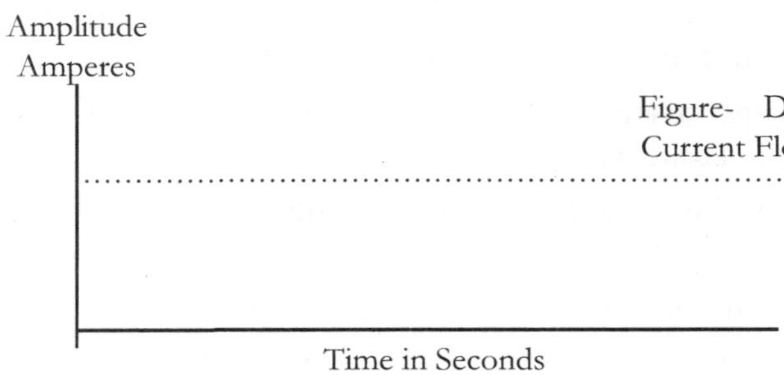

Figure- DC Current where the Current Flows in one direction only.

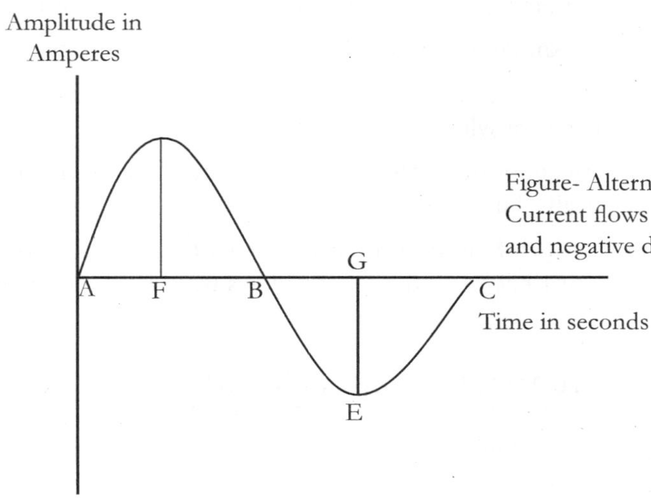

Figure- Alternating Current (AC Current flows in both positive and negative directions.

In AC current flows in both the directions forward an reverse. ADB is in positive direction whereas BEC is in negative directions.

ABD is positive cycle whereas BEF is negative cycle. In AC thus one complete cycle is made up of one negative and one positive cycle.

The Time to complete one complete Cycle is known as Time period.

Frequency is the number of cycles it completes in 1 Seconds. The unit of time period is Seconds and Frequency is Hertz. One cycle in one second is known as 1 Hertz.

Our domestic Power is AC Power and works at 50 Hertz meaning that it completes 50 Cycles in one 1 Seconds.

The maximum amplitude in one direction is the peak value (DF and GE) and the value from maximum to minimum value is Peak to Peak Value.(DE+GE)

RMS or Root Mean Square Value- This is the AC current value which will have the same heating effect as if a DC Current passing through a device will have.

The relation between Peak Value and RMS value is given by the Following equation:

$$Irms = \frac{I_o}{\sqrt{2}}$$

Where Irms is RMS value of the Current and Io is peak value of the current. Similar relationship holds for Voltage. In domestic Power we have 240 VAC. This means we have Vrms as 240 V whereas the Peak Voltage will be Vrms X √2= 340 V and the peak to peak will be 340 x2= 680 V.

INSULATORS, CONDUCTORS AND SEMICONDUCTOR

- **Conductors:** Materials that can allow current to flow through it freely are called good conductors or simply conductors. These materials have free electrons that can move easily and thus allow charge to flow freely.- Examples are Iron, Copper, Steel, Aluminium etc.
    - ❖ All metals are good conductor of electricity. However some metals are exceptions and are not good conductors of electricity and these are Manganese and Mercury.
- **Insulators:** Materials that do not allow current to flow through it freely are called bad conductors or simply Insulators. These materials do have free electrons that can move easily and thus do not allow charge to flow freely.- Examples are Wood, Earth, Plastic etc.
    - ❖ All non metals are bad conductor of electricity. However some non metals are exceptions and are good conductors of electricity. Graphite is one example of a non metal which is a good conductor of electricity.

- **Semiconductors: Semiconductors** are crystalline or amorphous solids which a resistance which is lower than have non metal but higher than metal for conducting electricity and their resistance decreases as their temperature increases, which is opposite to that of a metal. Finally, their conducting properties may be altered in useful ways by the deliberate, controlled introduction of impurities ("doping") into the crystal structure, which lowers its resistance but also permits the creation of semiconductor junctions between differently-doped regions of the extrinsic semiconductor crystal. The behaviour of charge carriers which include electrons, ions and electron holes at these junctions is the basis of diodes, transistors and all modern electronics.
    - ❖ Semiconductor devices can display a range of useful properties such as passing current more easily in one direction than the other, showing variable resistance, and sensitivity to light or heat. Because the electrical properties of a semiconductor material can be modified by doping, or by the application of electrical fields or light, devices made from semiconductors can be used for amplification, switching, and energy conversion.

Different Type of Electrical Cables and Their Specifications

Wire is a single electrical conductor, whereas a cable is a group of wires surrounded in sheathing.

Whether indoors or outdoors, proper wire and cable installation is of paramount importance - ensuring a smooth electricity supply, as well as passing electrical inspections. Each wire and cable needs to be installed carefully, from the fuse box to the outlets, fixtures and appliances. The Standard IS : 1255 - 1983 Indian Standard code for installation and maintenance of power cables up to and including 33 KV and national electrical rules and local building codes regulate the manner of installation and the types of wires and cables for various electrical applications.

Some factors that will affect your choice of electrical wiring include color, label information and applications. The information printed on the wire covering is all that you need to choose the correct wire for your home. Here's some detailed information on the various features of electrical wire, which will help you choose the correct composition:

1. **Size of Wires:** Each application requires a certain wire size for installation, and the right size for a specific application is determined by the wire gauge. Sizing of wire is done by the American wire gauge system. Common wire sizes are 10, 12 and 14 – a higher number means a smaller wire size, and affects the amount of power it can carry. For example, a low-voltage lamp cord with 10 Amps will require 18-gauge wire, while service panels or subpanels with 100 Amps will require 2-gauge wire.

2. **Wire Lettering:** The letters THHN, THWN, THW and XHHN represent the main insulation types of individual wires. These letters depict the following NEC requirements:
   - T – Thermoplastic insulation
   - H – Heat resistance
   - HH – High heat resistance (up to 194°F)
   - W – Suitable for wet locations
   - N – Nylon coating, resistant to damage by oil or gas
   - X – Synthetic polymer that is flame-resistant
3. **Types of Wires** – There are mainly 5 types of wire:
   - **Triplex Wires:** Triplex wires are usually used in single-phase service drop conductors, between the power pole and weather heads. They are composed of two insulated aluminum wires wrapped with a third bare wire which is used as a common neutral. The neutral is usually of a smaller gauge and grounded at both the electric meter and the transformer.
   - **Main Feeder Wires** Main power feeder wires are the wires that connect the service weather head to the house. They're made with stranded or solid THHN wire and the cable installed is 25% more than the load required.
   - **Panel Feed Wires:** Panel feed cables are generally black insulated THHN wire. These are used to power the main junction box and the circuit breaker panels. Just like main power feeder wires, the cables should be rated for 25% more than the actual load.
   - **Non-Metallic Sheathed Wires :** Non-metallic sheath wire, or Romex, is used in most homes and has 2-3 conductors, each with plastic insulation, and a bare ground wire. The individual wires are covered with another layer of non-metallic sheathing. Since it's relatively cheaper and available in ratings for 15, 20 and 20 amps, this type is preferred for in-house wiring.
   - **Single Strand Wires :** Single strand wire also uses THHN wire, though there are other variants. Each wire is separate and multiple wires can be drawn together through a pipe easily. Single strand wires are the most popular choice for layouts that use pipes to contain wires.
4. **Color Codes:** Different color wires serve different purposes, like:
   - Single phase Line- Brown.
   - Single phase Neutral- Blue.
   - Single Phase Earth- Green.
   - Three phase Line-L-1-Red
   - Three phase Line-L-2-Yellow

- Three phase Line-L-3-Blue
- Three phase Neutral-Black
- Three phase earth-Green/ Yellow strips on green

5. **Wire Gauge, Ampacity and Wattage Load:** To determine the correct wire, it is important to understand what ampacity and wattage a wire can carry per gauge. Wire gauge is the size of the wire, ampacity is how much electricity can flow through the wire and wattage is the load a wire can take, which is always mentioned on the appliances..

Understanding Electrical Cable

An electrical cable also has different types, color and application as its determining factors. Here's a brief about cables that you need to understand to determine the correct cable for your home.

1. **Types of Electrical Cables:** There are more than 20 different types of cables available today, designed for applications ranging from transmission to heavy industrial use. Some of the most commonly-used ones include:.
   - ***Non-Metallic Sheathed Cable:*** These cables are also known as non-metallic building wire or NM cables. They feature a flexible plastic jacket with two to four wires (TECK cables are covered with thermoplastic insulation) and a bare wire for grounding. Special varieties of this cable are used for underground or outdoor use, but NM-B and NM-C non-metallic sheathed cables are the most common form of indoor residential cabling.
   - ***Underground Feeder Cable:*** These cables are quite similar to NM cables, but instead of each wire being individually wrapped in thermoplastic, wires are grouped together and embedded in the flexible material. Available in a variety of gauge sizes, UF cables are often used for outdoor lighting and in-ground applications. Their high water-resistance makes them ideal for damp areas like gardens as well as open-to-air lamps, pumps, etc.
   - ***Metallic Sheathed Cable:*** Also known as armored or BX cables, metal-sheathed cables are often used to supply mains electricity or for large appliances. They feature three plain stranded copper wires (one wire for the current, one grounding wire and one neutral wire) that are insulated with cross-linked polyethylene, PVC bedding and a black PVC sheathing. BX cables with steel wire sheathing are often used for outdoor applications and high-stress installations.
   - ***Multi-Conductor Cable:*** This is a cable type that is commonly used in homes, since it is simple to use and well-insulated. Multi-conductor or multi-core (MC) cables feature more than one conductor, each of which is insulated individually. In addition, an outer insulation layer is added for extra security.

Different varieties are used in industries, like the audio multicore 'snake cable' used in the music industry.

- ***Coaxial Cable:*** A coaxial (sometimes heliax) cable features a tubular insulating layer that protects an inner conductor which is further surrounded by a tubular conducting shield, and might also feature an outer sheath for extra insulation. Called 'coaxial' since the two inner shields share the same geometric axis, these cables are normally used for carrying television signals and connecting video equipment.
- ***Unshielded Twisted Pair Cable:*** Like the name suggests, this type consists of two wires that are twisted together. The individual wires are not insulated, which makes this cable perfect for signal transmission and video applications. Since they are more affordable than coaxial or optical fiber cables, UTP cables are often used in telephones, security cameras and data networks. For indoor use, UTP cables with copper wires or solid copper cores are a popular choice, since they are flexible and can be easily bent for in-wall installation.
- ***Ribbon Cable:*** Ribbon cables are often used in computers and peripherals, with various conducting wires that run parallel to each other on a flat plane, leading to a visual resemblance to flat ribbons. These cables are quite flexible and can only handle low voltage applications.
- ***Direct-Buried Cable:*** Also known as DBCs, these cables are specially-designed coaxial or bundled fiber-optic cables, which do not require any added sheathing, insulation or piping before being buried underground. They feature a heavy metal core with many layers of banded metal sheathing, heavy rubber coverings, shock-absorbing gel and waterproof wrapped thread-fortified tape. High tolerance to temperature changes, moisture and other environmental factors makes them a popular choice for transmission or communication requirements.
- ***Twin-Lead Cable:*** These are flat two-wire cables that are used for transmission between an antenna and receiver, like TV and radio.
- ***Twinaxial Cable:*** This is a variant of coaxial cables, which features two inner conductors instead of one and is used for very-short-range high-speed signals.
- ***Paired Cable:*** With two individually insulated conductors, this cable is normally used in DC or low-frequency AC applications.
- ***Twisted Pair:*** This cable is similar to paired cables, but the inner insulated wires are twisted or intertwined.

2. **Cable Color Code:** Color coding of cable insulation is done to determine active, neutral and earth conductors. The NEC has not prescribed any color for phase/active conductors. Different countries/regions have different cable color coding, and it is essential to know what is applicable in your region. However, active conductors cannot be green/yellow, green, yellow, light blue or black..

- ***Cable Size:*** Cable size is the gauge of individual wires within the cable, such as 14, 12, 10 etc. – again, the bigger the number, the smaller the size. The number of wires follows the wire-gauge on a cable. So, 10/3 would indicate the presence of 3 wires of 10-gauge within the cable. Ground wire, if present, is not indicated by this number, and is represented by the letter 'G'.

Safety is very important, and if your installation of wires and cables is not proper, it could lead to accidents. Before you start any electrical project that includes wiring and cabling, you need to obtain permission from your local building inspector. Once the job is done, get the installation inspected for compliance with local codes and regulations.

Colour Codes

Function	*India Color Code (Old)*	*India Color Code (New)*
Single Phase Line		
Single Phase Neutral	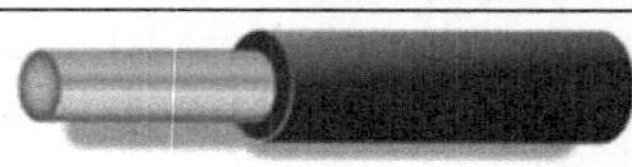	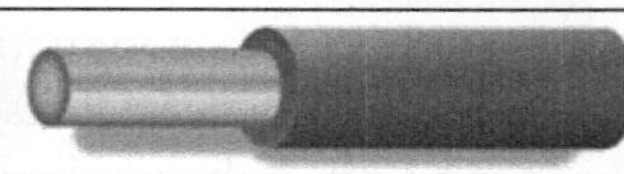
Single Phase Protective Ground or Earth		
Three Phase Line (L1)	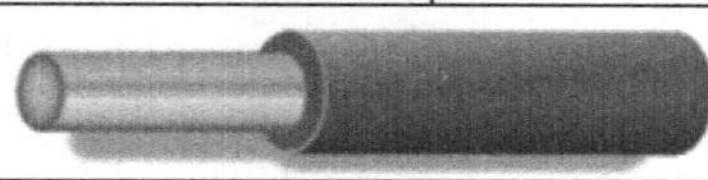	
Three Phase Line (L2)	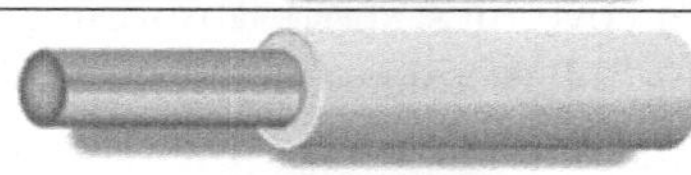	
Three Phase Line (L3)		
Three Phase Neutral (N)		
Three Phase Protective Earth or Ground (PE)	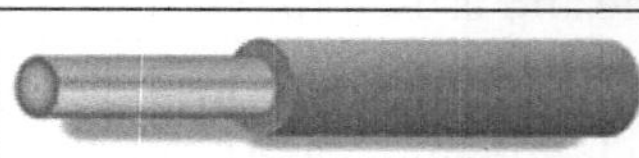	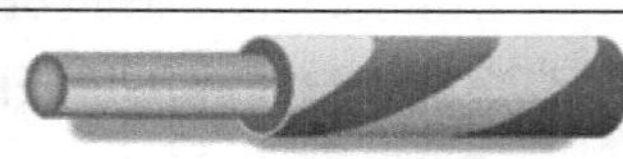

Basics of Electrical Safety and Practical procedures for electrical and personal safety measures.

If it is misused or poorly maintained, electrical equipment can be the cause of injury, death or fire. If it is well maintained, electrical equipment can save lives, improve the quality of lives and reduce capital expenditure. Electrical equipment and the electrical connections that supply power to it should always therefore be treated with respect and care.

Careful consideration should always be given to the placing of equipment. Damp conditions should be avoided and equipment should be positioned in a dry, clean, well ventilated area on a solid, level base. Equipment should be as near as possible to the electrical supply and extension leads should be discouraged.

Since most problems in this area occur with the plugs, sockets and cables supplying electrical power, the following measures would be important for electrical & personal safety:

1. **Socket outlets and plugs**
   - A convenient and safe socket outlet should be available.
   - Socket outlets should be at least 2 m from a sink or wash basin.
   - The socket outlet should be adequate for the electrical capacity for the equipment.
   - There should be proper grounding in the sockets.
   - Plugs should match the socket outlets.

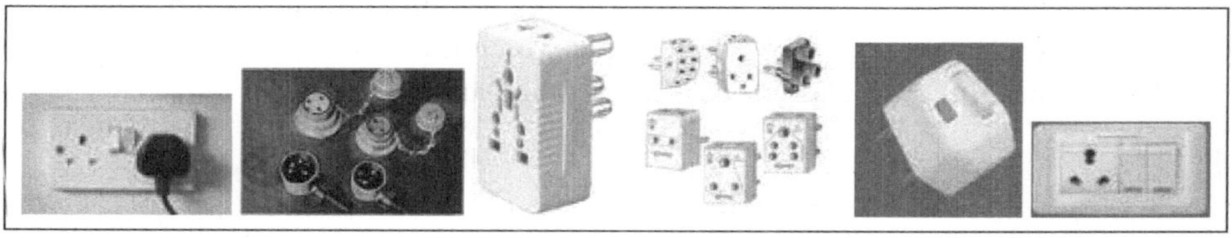

2. **Wiring of sockets and plugs:** The wiring of a plug is colour coded to help guard against electrical accidents. The colour codes in India as per Indian Electricity Rules are as follows and will help in identifying the way to connect the Phase, Neutral and Earth
   - ***Phase (or Live) – Red, Blue or Yellow***
     - ❑ This carries the electrical drive current from the supplier to the equipment. It is the most dangerous line. Only qualified staff should work with this.
   - ***Neutral – Black***
     - ❑ This returns the current to the supplier. It should not be connected to Earth.
   - ***Earth (or Ground) – Green OR Green with Yellow lines***
     - ❑ This is used for safety and protection. If equipment is housed in a metal case, the earth line will generally be connected to the case. The earth line in a socket is connected to a pipe or plate buried in the ground.

   Notes on earthing
   - The earthing will depend upon the type of equipment being used:

- ❑ If there are only two wires in the power cable, no earth connection is required
- ❑ If the cable fitted has three conductors then equipment needs to be earthed properly

❖ Always make sure that the earth wire is longer than the other two so that if the cable is accidentally pulled out of the plug, the earth wire is the last wire to become disconnected

3. **Sizes and types of sockets and plugs:** The current rating (i.e. the amount and size of equipment they can supply) is measured in Amperes, written 'A'. The rating and size of normally found plugs and sockets are:

❖ For low power operations 5 Amperes – small size

❖ For large power applications 15 Amperes – large size

Mains electricity comes at a specified voltage and is measured in Volts, written 'V'. The voltage in India is 220-240 V for single phase and 440 V for three phase operations. It also is delivered at a specific frequency, measured in Hertz, written 'Hz'. Mains electricity in India is at 50 Hz.

A variety of electrical plugs are found throughout India, so an adaptor plug set is recommended. Type D is most common, which is also known as the Old British Plug. It has three large round pins in a triangular configuration.

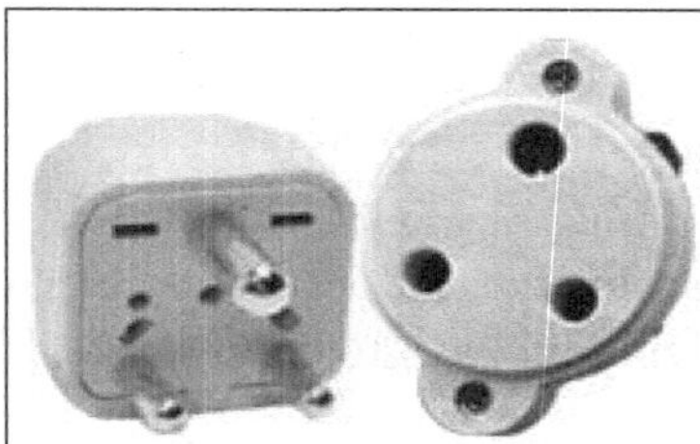

Type D Plug and Socket

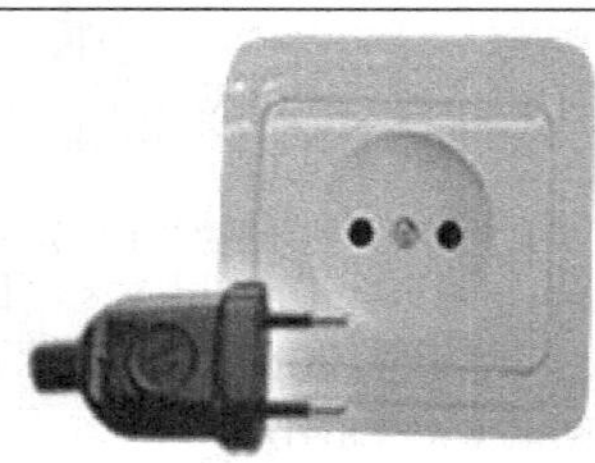

Type C Plug and Socket

The type C European 2-pin plug and electrical outlet is also very popular connector for common medical equipment which does not require earthing. Popularly known as the Europlug, it is used throughout continental Europe, parts of the Middle East, much of Africa, South America, central Asia, and the former Soviet republics.

4. **Mains cables:** Electricity is carried to the equipment through the mains cable: Points to be aware of are:

❖ No bare metal or internal coloured wire should be visible – the plastic insulation is there for safety

❖ Cable should not be repaired with insulating tape – water can still get inside

❖ Long flexible leads are dangerous – leads should be as short as possible

- ❖ The cable, plug and socket should never be allowed to get wet – water can conduct electricity

5. **Fuses:** Fuses are used as protection. If the current through them is greater than their specified rating, they blow. This breaks the circuit and stops the current, making the equipment safe. Points of safety regarding fuses are:
   - ❖ Always use the correct rating of fuse – voltage V (volts) and current A (amperes)
   - ❖ Always use the correct size of fuse – keep the old one to check against
   - ❖ NEVER replace the fuse with bare wire – it will not be safe
   - ❖ Circuit breakers are fuses that have buttons or switches for reset – they do not normally need replacing

Practical Exercise: Measuring wire sizes with wire guaze

STANDARD WIRE GAUGE

A standard wire gauge

British Standard Wire Gauge is a set of wire sizes given by BS 3737:1964 (now withdrawn), and is generally abbreviated to SWG. It is also known as: **Imperial Wire Gauge** or **British Standard Gauge**. Use of SWG sizes has fallen greatly in popularity, but is still used as a measure of thickness in guitar strings and some electrical wire. Cross sectional area in square millimetres is now a more popular size measurement. The current British Standard for metallic materials such as wire and sheet is BS 6722:1986, which is a solely metric standard.

The basis of the system is the mil, or 0.001 in. No. 7/0, the largest size, is 0.50 in. (500 mils or 12.7 mm) in diameter, and the smallest, No. 50, is 0.001 in. (1 mil or about 25 μm) in diameter. Between each gauge, the weight diminishes by approximately 20%. Because the weight per unit length is related to the area, and therefore the square of the diameter, the diameter diminishes by approximately 10.6%:

A table of wire gauges and diameters is shown below. The relationship of diameter to gauge is piecewise linear, only approximating a (constant-ratio) exponential curve.

British SWG (Standard Wire Gauge) diameters

SWG	in	Mm	Step
7/0	0.500	12.700	0.036"/gauge
6/0	0.464	11.786	0.032"/gauge
5/0	0.432	10.973	
4/0	0.400	10.160	0.028"/gauge
3/0	0.372	9.449	0.024"/gauge
2/0	0.348	8.839	
0	0.324	8.230	
1	0.300	7.620	
2	0.276	7.010	
3	0.252	6.401	0.020"/gauge
4	0.232	5.893	
5	0.212	5.385	
6	0.192	4.877	0.016"/gauge
7	0.176	4.470	
8	0.160	4.064	
9	0.144	3.658	
10	0.128	3.251	0.012"/gauge
11	0.116	2.946	
12	0.104	2.642	
13	0.092	2.337	
14	0.080	2.032	0.008"/gauge
15	0.072	1.829	
16	0.064	1.626	
17	0.056	1.422	
18	0.048	1.219	
19	0.040	1.016	0.004"/gauge
20	0.036	0.914	
21	0.032	0.813	
22	0.028	0.711	

23	0.024	0.610	0.002"/gauge
24	0.022	0.559	
25	0.020	0.5080	
26	0.018	0.4572	0.0016"/gauge
27	0.0164	0.4166	
28	0.0148	0.3759	0.0012"/gauge
29	0.0136	0.3454	
30	0.0124	0.3150	0.0008"/gauge
31	0.0116	0.2946	
32	0.0108	0.2743	
33	0.0100	0.2540	
34	0.0092	0.2337	
35	0.0084	0.2134	
36	0.0076	0.1930	
37	0.0068	0.1727	
38	0.0060	0.1524	
39	0.0052	0.1321	0.0004"/gauge
40	0.0048	0.1219	
41	0.0044	0.1118	
42	0.004	0.1016	
43	0.0036	0.0914	
44	0.0032	0.0813	
45	0.0028	0.0711	
46	0.0024	0.0610	
47	0.0020	0.0508	
48	0.0016	0.0406	
49	0.0012	0.0305	0.0002"/gauge
50	0.0010	0.0254	

Practical Exercise: How to Identify the Live, Neutral and Earth on Power Socket

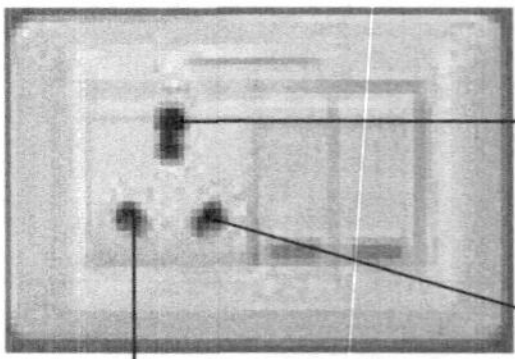

If the wires are not connected as above you need to open the Power Socket and connect the wires as indicated.

Practical Exercise: How to use a Tester to Monitor AC Power

Monitoring AC Power is simple with the Testers. The popular tester looks as below. Insert the tip insi

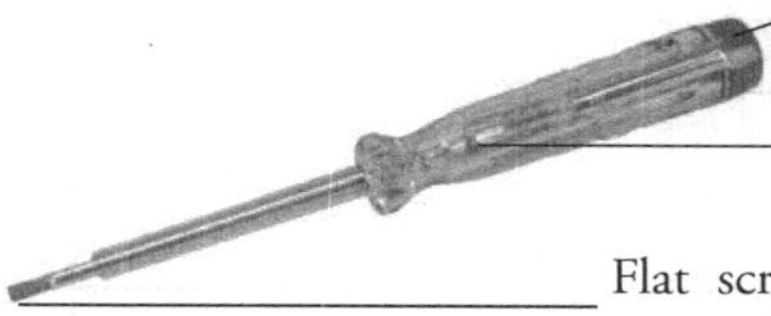

Insert the Flat tip in the Socket and put the finger on the top of the tester. The Lamp will light if the Power is there when it touches the LIVE wire. If the LIVE wire is not in the LEFT side open the socket and connect the LIVE to the left and Neutral to the RIGHT. The Lamp will not glow in the Neutral or the Earth.

Practical Exercise: Construct a Test Lamp and use it to check the Mains healthiness.

A simple Test Lamp is constructed as below:

1. Take a holder and connect a two wires to the holder. Make sure the wires are well insulated.
2. Connect the other end to two probes and make sure that the point where connection is being done is well insulated.
3. Connect a Lamp say 30 W bulb.
4. Insert the probe between LIVE and NEUTRAL Socket . The Lamp should light if the POWER is there.

5. Insert the probe between EARTH and NEUTRAL. The Lamp should not light if the NEUTRAL AND GOUND are OK.
6. Insert the probe between the EARTH and LIVE points. The Lamp should light if the EARTH and LIVE is ok and if POWER is there.

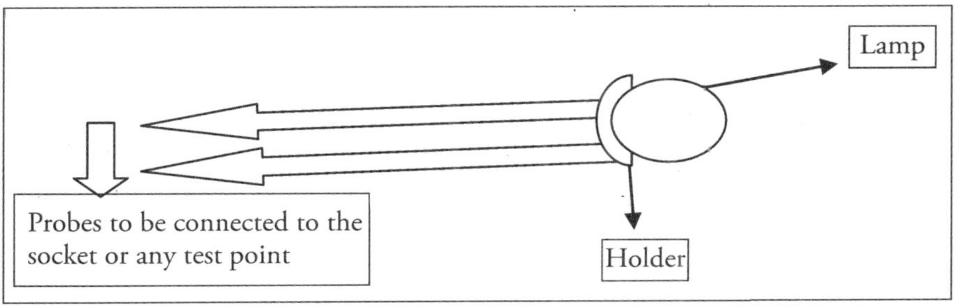

If on testing a socket the Neutral is found on the LEFT open the socket and connect the LIVE to the LEFT and NEUTRAL to the RIGHT hand side in the socket.

Practical Exercise: Measure the Voltage between Neutral and Ground and rectify the Earthing

Ground Or Earthing protects us from the fatal effect in case a device gets short circuited. The Live current will be carried to the Earth and will not give us hazardous shocks in case the body touches the LIVE. Neutral is meant to complete the CIRCUIT with Live wire.

Measure the Voltage between Ground and Neutral and make sure it is LOW say less than 5 V. If it is more than that we have to make a proper ground.

To make a proper ground dig up the earth around two feet and put some charcoal , salt and water into it. Then put a Copper rod inside it and fill up the earth. Attach an Earth wire to the rod and attach the other end of the wire to the Ground connection on the socket or to the Mains earth connection.

Occasionally keep watering it for good Earthing.

Practical Exercise: Making the mains cable termination

We need to find the best way to terminate the ends of a cable or wire. The important common termination options are provided below:

Wire Termination Options

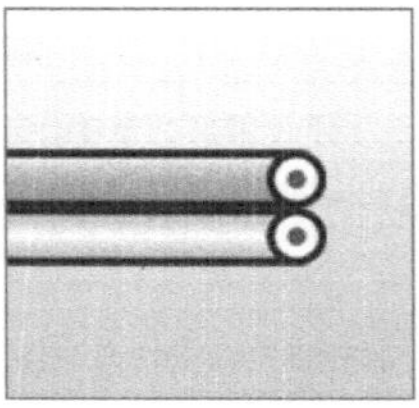	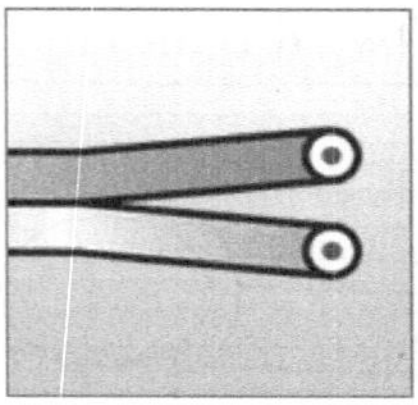	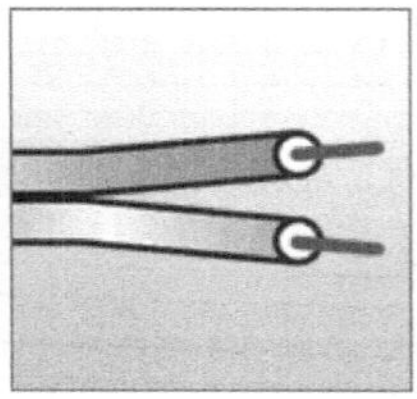	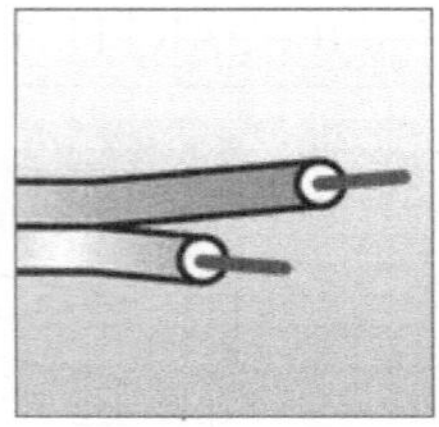
End Cut	Wire Slit	Wire Strip	Cut Back
Wire is cut even at customer specified length.	Where parallel wire is used, ends can be separated from each other to customer specifications.	Wire ends are "stripped" of their insulation material to expose the inner copper strands.	Wires are cut and terminated at different lengths according to customer specifications

Cable Termination Options

	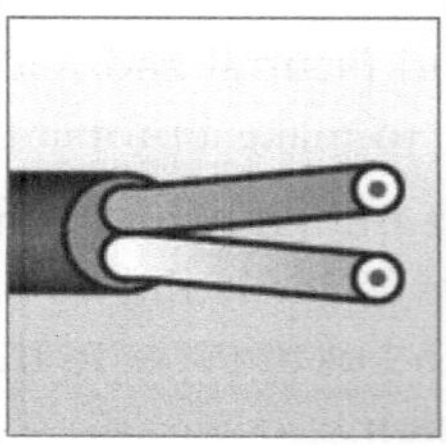	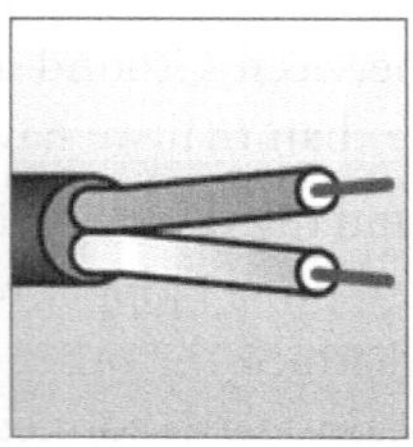	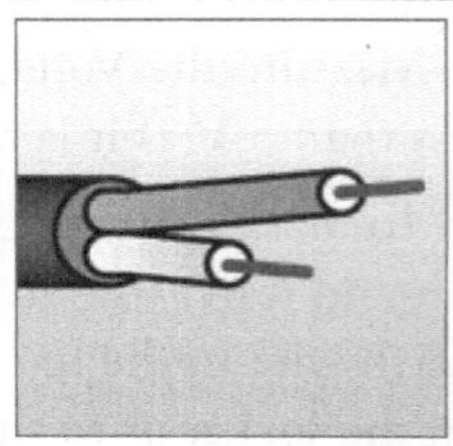
Blunt Cut	Remove outer-Jacket	ROJ & Stripped Wires	Cutback Wires
Cable is cut even at customer specified length.	Also known as R.O.J., the outer insulating cable jacket is removed to expose the inner wires. Length is determined by customer.	Outer cable jacket is removed and wire ends are stripped of insulating material.	Outer cable jacket is removed and wires are cut and terminated at different lengths.

Termination Connectors

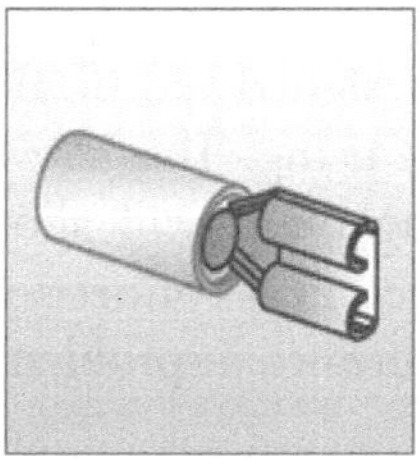	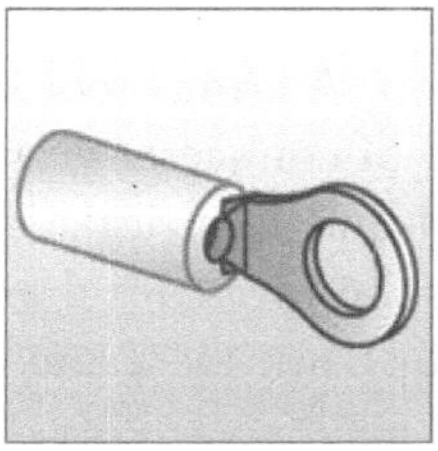	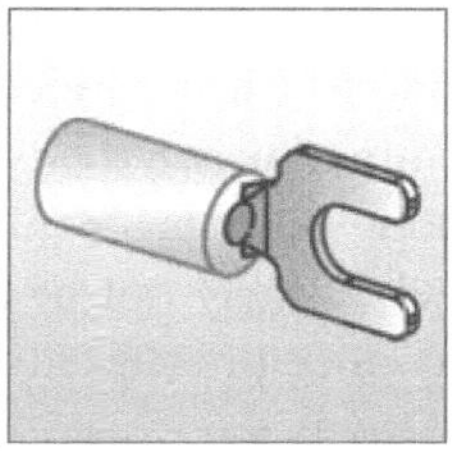	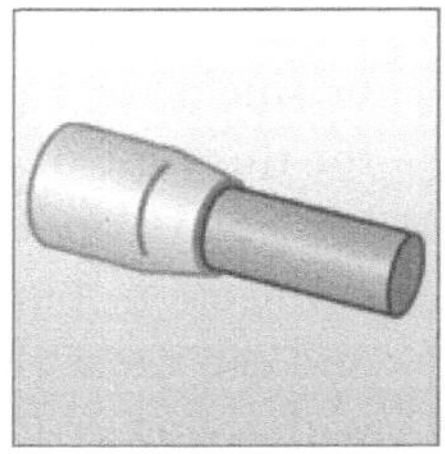
Quick connect	Ring Terminal	Spade or Fork Terminal	Ferrule Termina
Available In: Un-Insulated, Pre Insulated, Fully Insulated, Right-Angle (Flag)	Used to attach product with screw or bolt. Perfect for a ground connection. Also available in locking.	Similar to a Ring Terminal, but removable without completely unscrewing bolt or screw	For use with Terminal Blocks. Used to crimp and secure bare copper strands. Also available in a double for crimping 2 separate wires.

Termination Types

1. Solder Type A solder type connection allows for a strong, solid mechanical and electrical connection. Clean the connection well. For electrical circuits you must use a rosin type flux to clean all connections. Do not use acid flux that is commonly used for plumbing installation. The acid based flux will cause corrosion and inherently cause intermittent problems with the electrical signal. The choice of solder is also important. Using a solder standard 60/40 formula will meet the majority of your soldering needs. However, lead-free and high-grade silver solder is available for special applications. Also, use a soldering iron of the proper wattage. If the soldering iron is not hot enough, you may not be able heat the connection enough to get a good solder joint. This may cause what is known as a "cold" solder joint and can cause intermittent problems like opens to occur. However, if the soldering iron is too hot, you can cause damage to the components of the system near the connection. This can also cause the insulation to possibly melt causing the bare primaries to make contact with each other resulting in a short.
2. **Crimp Type:** A crimp type connection allows for quick and simple installation while still maintaining a mechanical and electrical connection fairly close to a solder type termination. Solid or stranded wire can be used in this type of termination. Some of the key points to remember for a good clean connection are as follows: 1. Make sure you use the proper size connector for the type of cable you are using. 2. Make sure all of your cuts and stripping are clean. 3. Avoid

nicks as much as possible. 4. Use the proper crimp tool, don't try to improvise with pliers, etc. The most common crimp method involves two crimps, one on the insulation for a stronger mechanical connection and one on the conductor or shield for a good electrical connection. A crimp tool is designed specifically for this type of termination for the type of connector you are using. This allows for good connections both mechanical and electrical. Using pliers will allow connection. However, it may not be a solid mechanical or electrical connection and can cause the connector to eventually come loose and intermittent problems with the electrical signal can occur.

3. **Insulation Displacement:** This type of termination is usually used in punch down blocks, wall connectors, and in the back of patch panels. This type of termination eliminates the need for stripping the conductor insulation. As the conductor is pushed though the clip, the insulation is cut into and the metal clip contact makes contact with the wire. The best type of wire to use is a solid conductor. If you use a stranded conductor the force of the termination may allow the clip to cut some of the strands. Also, stranded wire will "crush" somewhat which will not allow for a solid connection. However, both types of conductors are used in various systems. The connection is made by using a punch down tool. Some patch panel manufacturers supply a termination cap that terminates several wires at once without the use of a tool. Some tools will allow use of different bits for use with various terminations. (i.e. 66 block, 110 block, etc.). Just like the crimping type termination, it is important to use the correct bit for the type of termination you are doing.
4. **Direct Connection (Utility Block/ Screw Terminals):** This type of termination has several names. Utility block, barrier strip, or screw terminals are just a few. This type of termination can use either solid or stranded conductors. It allows for easy termination as well as quick changing of wire in the future. The main point to remember about a screw connection is to strip back the insulation only to the amount of conductor that will wrap around the screw and to place the wire in the same direction as the screw turns when tightening. This will "pull" the wire in tighter as the screw tightens. If you wrap the wire around the screw opposing the tightening rotation of the screw, the wire will be pulled outward and will become unwrapped around the screw.

CHAPTER

2

Multimeter

Analogue Multimeter, (VOM or VOA)

The analog or analogue multimeter often called a VOA meter for measuring Voltages in Volts, **Current in** Amps and **Resistance in** Ohms measurements or a VOM - Volts Ohm Meter.

Principle of operation of Analog Multi-meter - The analogue VOA meter or VOM is based around the moving coil galvanometer.

The meters will have a full scale deflection, FSD, i.e. the maximum reading calibrated on the scale of a certain current. Typically this might be a value of 50μA. The functionality of the basic meter is then extended by adding series and shunt resistors to enable voltage and current to be measured.

- ***Extending range of a meter for current measurements:*** To extend the current ranges of a basic analogue meter, a resistor is placed in parallel with the meter (This is known as Shunt Resistor). In this way the shunt resistor takes current and for the same overall current flowing through the meter, more can flow through the overall circuit.

  How to calculate the value of shunt resistor- The value of the shunt resistor is easy to calculate using Ohm›s law. Using this it can be determined that the proportion of the current flowing in each leg is inversely proportional to the resistance. This if the moving coil meter has a full scale deflection of 50 μA and a resistance of 2 kΩ for a 1mA FSD, 0.95 mA needs to flow in the shunt resistor for the same voltage across the shunt resistor and the meter itself. Therefore the resistance of the shunt resistor needs to be: 5 / 95 x 2 kΩ = 105.3 Ω.

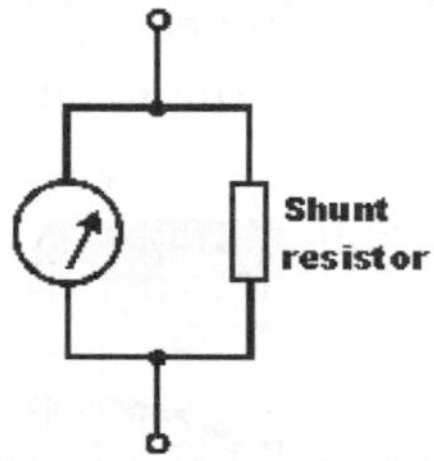

Analogue amp meter using a shunt resistor for higher current capability

- ***Extending the range for voltage measurements:*** For voltage measurements, resistors are placed in series with the meter.

Analogue voltmeter using a moving coil meter

It is easy to calculate the value for the resistor. Knowing the resistance of the moving coil meter and its full scale deflection, it is possible to use Ohm's law to calculate the required values.

For example take a moving coil meter with a 50 μA FSD and a coil resistance of 2kΩ. For a voltage of 10 volts to enable 50μA to flow the total resistance must be V/I = R or 10 / 50 x 1^{-6} = 200 kΩ. Thus the series resistor required is 200 kΩ - 2 kΩ i.e. 198 kΩ.

- **Resistance capability for a VOA meter:** In order to provide the resistance measurement capability, an additional battery is required. This provides a current source to drive current through the external resistor. The amount of current flowing provides an indication of the resistance.

  When making resistance measurements using an analogue multimeter, it is found that the high resistance indications are at the left hand section of the meter, i.e. when less current is flowing, and the low values of resistance are indicated at the right hand end of the meter scale, because a higher current flows. This may be a little confusing at first, but one quickly becomes accustomed to this. When using a resistance measurement on an analogue multimeter or analogue VOA meter, it is first necessary to «zero» the meter. This is needed to calibrate out any variations in the battery voltage. It is achieved simply by sorting out the two analogue multimeter probes and adjusting the control normally labelled «Zero» for zero ohms. Once this has been achieved the meter can be used accurately. A further point to note is that the negative terminal of the analogue multimeter is positive to the positive terminal, i.e. the polarity on the terminals is the opposite of what might normally be expected. For most measurements this is not of any consequence, although for some measurements of semiconductors it will have a bearing.

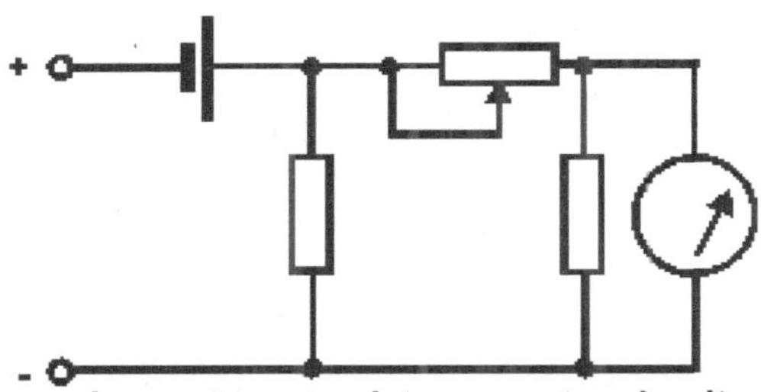

Analogue ohm-meter or resistance meter circuit

It can be seen that by adding the shunt and series resistors as well as a resistor network and battery, for resistance, it is possible to provide a considerable amount of additional capability for the basic analogue moving coil meter.

STUDY OF DIFFERENT CONTROLS ON MULTIMETER

Making measurements using an analogue multimeter

When using an analogue multimeter, measurements for voltage, current and resistance need to be made in different ways. To illustrate the way in which these different tests can be made using an analogue multimeter, the simple circuit shown below will be used:

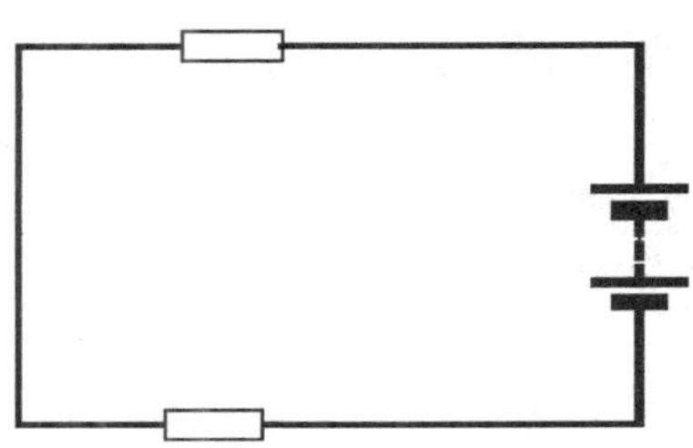

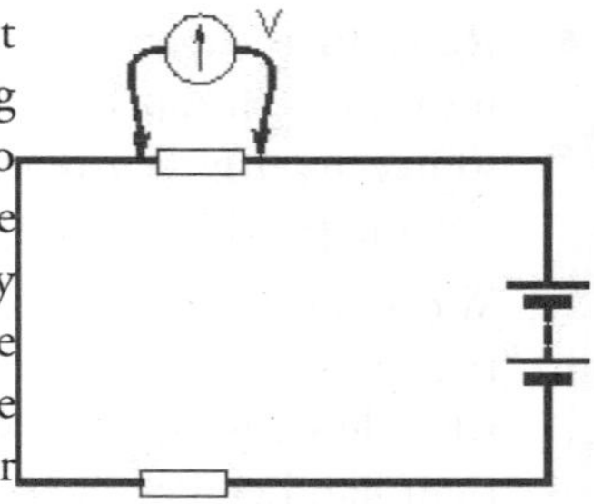

1 ***Voltage measurements:*** The voltage measurement is the easiest form of measurement to make using an analogue multimeter. The two leads need to be connected across the area of the circuit where the voltage measurement is to be made. Typically the "Common" or "COM" connection on the multimeter is used for the negative voltage end of the measurement and the connection marked "Volts" or similar goes to the positive end of the measurement. For some high or low voltage measurements, there may be a separate connection available and this should be chosen.

Once the relevant connections of the multimeter have been chosen Select the "VOLTAGE" and the switch can be turned to the correct range and the measurement taken.

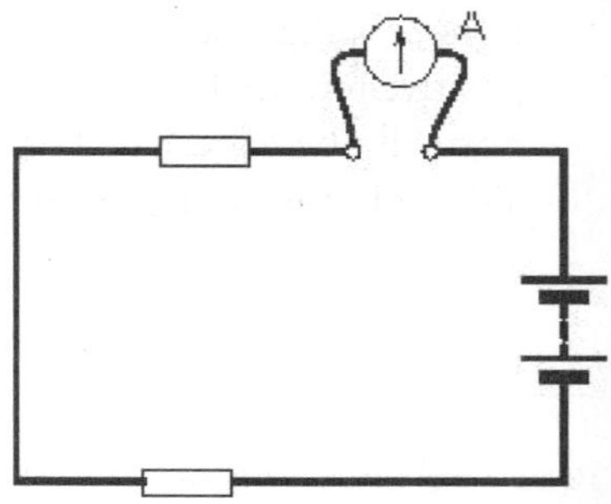

- ***Current measurements:*** When using a multimeter to make a current measurement, the meter is placed in series with the circuit where the current measurement is to be made. The " Current" mode is selected and the switch is selected to the desired range.
- ***Resistance measurements:*** For any resistance measurement, the component to be measured should be removed from the circuit as other paths will be present that will distort the reading. Also any power remaining within the circuit will add to the inaccuracies. Never measure resistance when the circuit is powered, as power from the circuit will not only distort the reading, but could damage the meter.

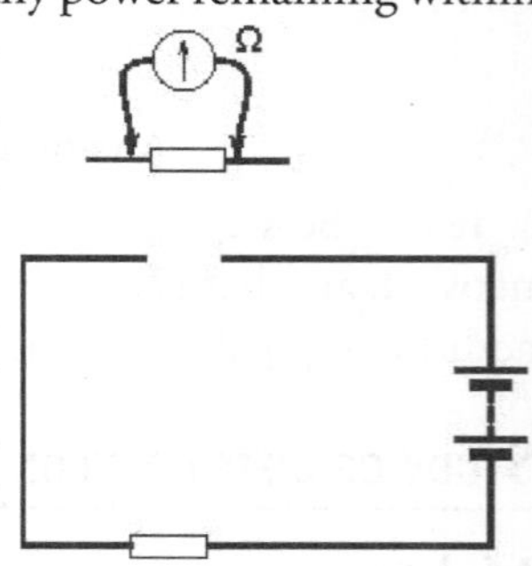

Using analogue test meters can be every bit as easy as using digital multimeters. Their accuracy is normally more than adequate and with a little care and understanding in their use they provide an excellent multimeter.

Precautions in Using the Analog Multimeters

Parallax errors

One cause of errors on an analogue multimeter, or any analog meter for that matter is parallax errors. This is an important concept when using an analogue multimeter.

When viewing the meter, the eye should be at right angles to the plane of the meter back markings. In this way there is no error from viewing the needle at an angle.

Some high end professional meters such as the AVO have a mirror in the scale. In this way it is possible to assess whether the eye is directly in front of the scale - when the eye is viewing correctly, it will not be possible to see the reflection as it is masked out by the needle itself. The offset view below indicates this.

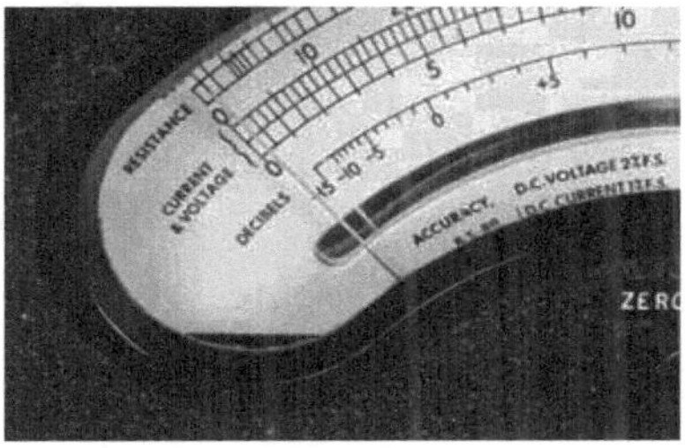

In addition to the mirror, often the needle is made thin in the place of the scale, but much larger in the plane at right angles to the scale. In this way it has sufficient strength and it is also does not take much meter scale area.

Using the correct range

Another concept in knowing how to use an analogue multimeter is that of knowing which range to use.

In terms of the view of the meter, the best accuracy is gained when the meter is towards the full scale deflection, FSD. In this way a given percentage change in the reading gives the maximum and hence most visible change in meter deflection, and accordingly the most accurate reading.

However care has to be taken not to overload the meter by placing it on a range much too low for the reading to be taken. If this occurs the meter can swiftly move to the end-stop, and damage may occur if it is overloaded too much. It is always best to start well below the range expected to give full scale deflection and switch the range when everything has settled.

Practical Exercise: How to make Analogue meter zero position adjustment

Analogue multimeters, and in fact any analogue moving coil meter will have a zero adjuster.

This zero adjustment should not normally need to be touched, but it may vary slightly with time and temperature.

Adjustment should be made with the meter not in use and it should be gently adjusted with a screwdriver to ensure the meter is properly zeroed. Care should be taken to ensure that the eye is directly above the meter, and any mirror in the meter should be used to ensure that accuracy is maintained.

The adjustment should be undertaken with the meter level as if the meter is placed horizontally, for example, the zero position will change.

0Ω Adjustment

When using a multimeter for resistance measurements it is necessary to adjust the zero ohm position. This is required to compensate for a number of variations from small tolerances in the components in the meter to the state of the battery within the meter that is used to supply the current needed for the measurement.

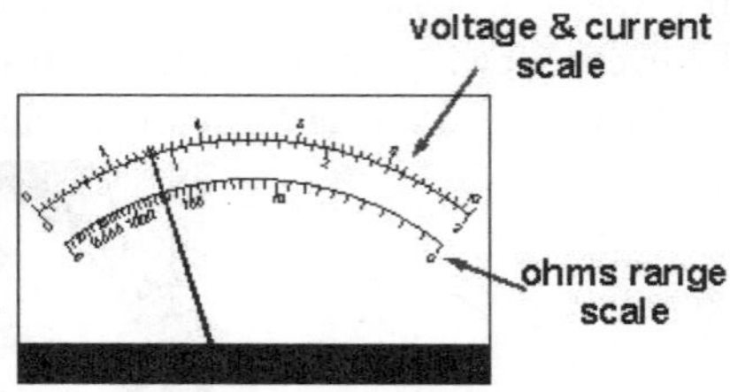

Scale on a typical analogue multimeter

To undertake the ohm-meter zeroing, the test probes for the meter should be shorted together to give a zero ohm resistance between the terminals of the meter, and the small "Ohms Adjustment" control should be used to give full scale deflection on the meter that corresponds to the zero ohms position.

Note that on an analogue meter zero ohms corresponds to full scale deflection, and increasing resistance gives a lower level of deflection. In this way the meter scale is reversed - higher values of resistance are to the left of the scale and lower ones to the right.

Finishing using multimeter

When the measurements using the analogue multimeter have been completed, it is always wise to return he meter to its off position. This prevents the meter being picked up and connected for a voltage measurement, when, for example it has been left set for a current measurement. In this case excess current could flow through the meter causing damage to the meter and also with the possibility of damage to the circuit under test.

If no specific "Off" position is available, after use the multimeter should be switched to the highest voltage range.

Digital Multimeters

Familiarization with operation controls of digital Multi-meter Principle of operation of digital Multi-meter

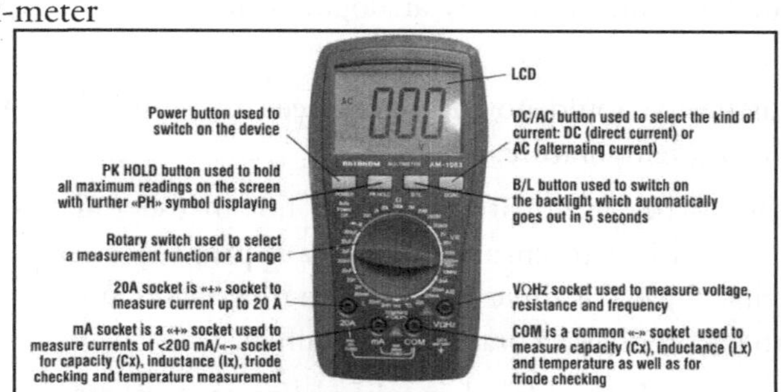

Figure 2.1: Operational controls on a digital multimeter

The digital multimeter, DMM, is one of the most common items of test equipment used in the electronics industry today.

While there are many other items of test equipment that are available, the multimeter is able to provide excellent readings of the basic measurements of amps, volts and ohms.

In addition to this the fact that these digital multimeters use digital and logic technology, means that the use of integrated circuits rather than analogue techniques, enables many new test features to be embedded in the design.

As a result, most of today's digital multimeters incorporate many additional measurements that can be made.

DMM facilities

While the facilities that a digital multimeter can offer are much greater than their analogue predecessors, the cost of DMMs is relatively low. DMMs are able to offer as standard the basic measurements that would typically include:

- Current (DC)
- Current (AC)
- Voltage (DC)
- Voltage (AC)
- Resistance

However, using integrated circuit technology, most DMMs are able to offer additional test capabilities. These may include some of the following:

- Capacitance
- Temperature
- Frequency
- Transistor test - hfe, etc
- Continuity (buzzer)

While some of these additional test features may not be as accurate as those supplied by dedicated test instruments, they are nevertheless very useful, especially where approximate readings only are needed.

In addition to an increase in the number of basic measurements that can be made, refinements of some of the basic measurements are also available on some models. True RMS multimeters are available. In many instances, AC waveforms use forms of average measurements that are then converted to RMS measurements using a form factor. This method of measurement is very dependent upon the shape of the waveform and as a result a true RMS digital multimeter may be required. In addition to the availability of a true RMS meters, similar refinements of the other basic measurements are also available in some instances.

In addition to the additional measurement capabilities, DMMs also offer flexibility in the way measurements are made. Again this is achieved because of the additional capabilities provided by the digital electronics circuitry contained within the digital multimeter. Many instruments will offer two additional capabilities:

- ***Auto-range:*** This facility enables the correct range of the digital multimeter to be selected so that the most significant digits are shown, i.e. a four-digit DMM would automatically select an appropriate range to display 1.234 mV instead of 0.012 V. Additionally it also prevent overloading, by ensuring that a volts range is selected instead of a millivolts range. Digital multimeters that incorporate an auto-range facility usually include a facility to 'freeze' the meter to a particular range. This prevents a measurement that might be on the border between two ranges causing the meter to frequently change its range which can be very distracting.
- ***Auto-polarity:*** This is a very convenient facility that comes into action for direct current and voltage readings. It shows if the voltage of current being measured is positive (i.e. it is in the same sense as the meter connections) or negative (i.e. opposite polarity to meter connections). Analogue meters did not have this facility and the meter would deflect backwards and the meter leads would have to be reversed to correctly take the reading.

DMM measurements

Digital multimeters can make a variety of different measurements. Using them is normally very easy, but the right techniques must be adopted to make the measurements correctly.

This DMM tutorial includes pages outlining the techniques required for making various measurements.

- ***How to use a DMM:*** There are some basic hints and tips on using a DMM that apply across most DMM measurements.
- ***Voltage measurements:*** The voltage measurement is one of the most widely used and most simple DMM measurements.

Practical Exercise: How to use a digital multmeter for AC and DC voltage measurement

One of the most widely used measurements for a DMM is that of voltage. Knowing how to use a digital multimeter to make the best voltage measurements can ensure the most accurate measurements are made.

When making a voltage measurement the probes from the digital multimeter are placed across the points where the voltage is to be measured.

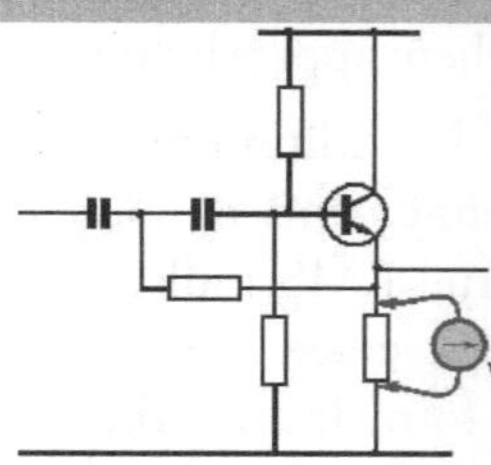

A typical digital multimeter voltage measurement

- ***Current measurement:*** Current measurements are a

little more complicated requiring the digital multimeter to be placed within the circuit

- ***How to use a digital multmeter for current:*** When making a current measurement, the current flowing within the circuit needs to be detected. The traditional method of achieving this is to break the circuit and place the digital multimeter acting as a current meter in circuit. In this way the current flows through the meter. the level of current can then be detected and displayed.

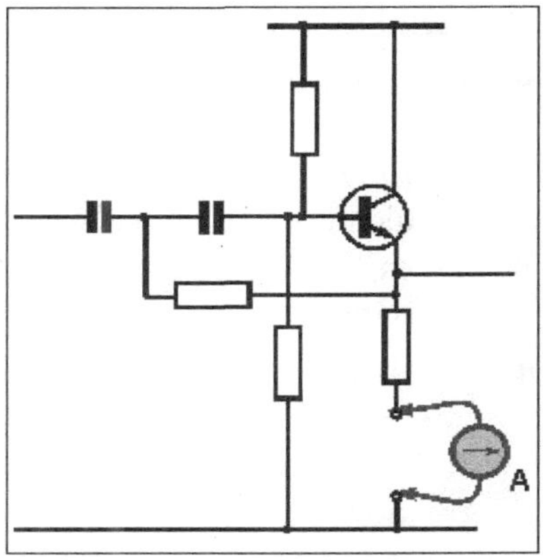

Figure 2.2: A typical digital multimeter current measurement

- ***Resistance measurement:*** It is often necessary to measure the resistance of a component. This is easy to accomplish, but certain precautions must be observed.
- ***How to use a digital multmeter for resistance:*** When making a resistance measurement with a digital multimeter, it is necessary to place the item under test across the probes of the multimeter. It must be removed from the circuit, otherwise stray conduction paths will be present. Even if measuring continuity the unit must be off and power removed. Should any voltages be present, these will at the very least distort the readings, and at worst they could damage or even destroy the instrument. Care must be taken.

Using a DMM for other measurements

In addition to the basic voltage, current and resistance measurements, many digital multimeters offer a range of other measurements that they can make. The actual measurements depend upon the actual DMM being used. However knowing how to use the digital multimeter to make these measurements means that the results can be assessed within the capabilities of the instrument.

Other measurements that may be incorporated into digital multimeters may include:

- ***Continuity sounder:*** This is one of the most common additions found in a digital multimeter. This capability is sometimes included on a low value Ohms range, or may have its own switch position. The idea is that it is possible to test for continuity and listen for a buzz, rather than having to continually turn away from looking at the unit under test. When using the continuity tester, the same precautions that should be implemented when using the ordinary resistance ranges should be observed, especially the unit should not be powered up..
- ***Frequency:*** Some digital multimeters can be used to measure frequency. This is one of the less commonly included ranges, but can be used to give a rough

indication of frequency up to a few hundred kilohertz. Normally the ranges do not extend very high, and they are not normally very accurate. However they are useful for some low frequency measurements, where accuracy is not paramount.

- ***Capacitance:*** With some digital multimeters, it is possible to use them to measure capacitance. Again this facility is not available on all meters, but some have the ability to measure it. Like resistance measurements, capacitance measurements should ideally be undertaken on the component when it is not in circuit. Normally the capacitance ranges are limited, and they are not able to measure small levels of capacitance.
- ***Transistor tester:*** Occasionally multimeters have the facility to measure transistor parameters. In particular they measure the He or B of the transistor. Typically there are three connections for each of the PNP and NPN varieties, so you need to know what sort it is before testing.

Digital multimeters are widely used and very useful items of test equipment. They enable measurements of quantities such as current, voltage and resistance to be made very quickly and easily. In addition to this, many DMMs are able to measure other useful parameters, making these items even more useful. While they do not allow more complicated measurements to be made, if many engineers were allowed only one item of test equipment, it would probably be the digital multimeter.

Basics of how a DMM works

The key process that occurs within a digital multimeter for any measurement that takes place is that of voltage measurement. All other measurements are derived from this basic measurement.

Accordingly the key to understanding how a digital multimeter works is in understanding this process.

There are many forms of analogue to digital converter, ADC. However the one that is most widely used in digital multimeters, DMMs is known as the successive approximation register or SAR. Some SAR ADCs may only have resolution levels of 12 bits, but those used in test equipment including DMMs generally have 16 bits or possibly more dependent upon the application. Typically for DMMs resolution levels of 16 bits are generally used, with speeds of 100k samples per second. These levels of speed are more than adequate for most DMM applications, where high levels of speed are not normally required.

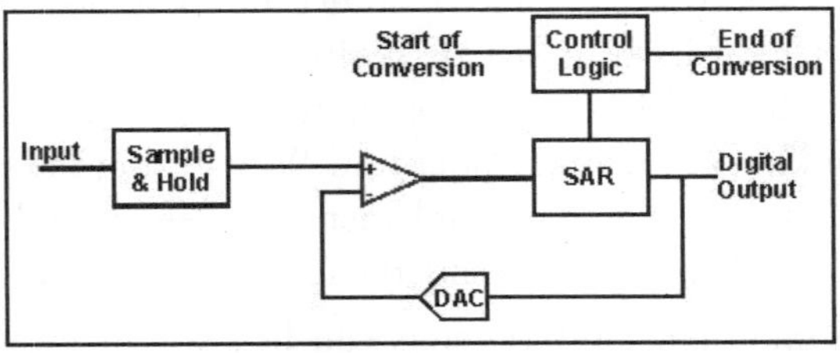

Figure 2.3: Successive approximation register ADC used in most DMMs

As the name implies, the successive approximation register ADC operates by successively homing in on the value of the incoming voltage.

The first stage of the process is for the sample and hold circuit to sample the voltage at the input of the DMM and then to hold it steady.

With a steady input voltage the register starts at half its full scale value. This would typically require the most significant bit, MSB set to "1" and all the remaining ones set to "0". Assuming that the input voltage could be anywhere in the range, the mid-range means that the ADC is set in the middle of the range and this provides a faster settling time. As it only has to move a maximum of the full scale rather than possibly 100%.

To see how it works take the simple example of an 4-bit SAR. Its output will start at 1000. If the voltage is less than half the maximum capability the comparator output will be low and that will force the register to a level of 0100. If the voltage is above this, the register will move to 0110, and so forth until it homes in on the nearest value.

It can be seen that SAR converters, need one approximating cycle for each output bit, i.e. an n-bit ADC will require n cycles.

DMM operation

Although the analogue to digital converter forms the key element within the instrument, in order to fully understand how a digital multimeter works, it is necessary to look at some of the other functions around the ADC.

Although the ADC will take very many samples the overall digital multimeter will not display or return every sample taken. Instead the samples are buffered and averaged to achieve high accuracy and resolution. This will overcome the effects of small variations such as noise, etc., noise created by the analogue first stages of the DMM being an important factor that needs to be overcome to achieve the highest accuracy.

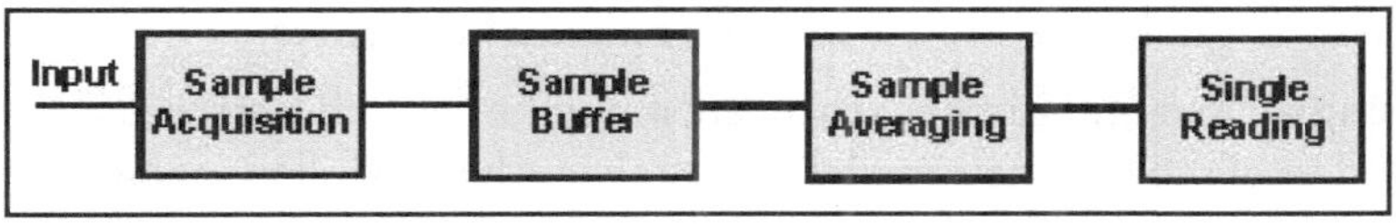

Figure 2.4: Operation flow diagram for operation of a DMM

Measurement time

One of the key areas of understanding how a digital multimeter works is related to the measurement time. Apart from the basic measurement there are a number of other functions that are required and these all take time. Accordingly the measurement time of a digital multimeter, DMM, may not always appear straightforward.

The overall measurement time for a DMM is made up from several phases where different activities occur:

- ***Switch time:*** The switch time is the time required for the instrument to settle after the input has been switched. This includes the time to settle after a measurement type has been changed, e.g. from voltage to resistance, etc. It also includes time to settle after the range has been changed. If auto-ranging is included the meter will need to settle if a range change is required.
- ***Settling time:*** Once the value to be measured has been applied to the input, a certain time will be required for it to settle. This will overcome any input capacitance levels when high impedance tests are made, or generally for the circuit and instrument to settle.
- ***Signal measurement time:*** This is the basic time required to make the measurement itself. For AC measurements, the frequency of operation must be taken into account because the minimum signal measurement time is based on the minimum frequency required of the measurement. For example, for a minimum frequency of 50 Hz, an aperture of four time the period is required, i.e. 80 ms for a 50Hz signal, or 67ms for a 60Hz signal, etc.
- ***Auto-zero time:*** When autorange is selected, or range changes are made, it is necessary to zero the meter to ensure accuracy. Once the correct range is selected, the auto-zero is performance for that range.
- ***ADC calibration time:*** In some DMMs a calibration is periodically performed. This must be accounted for, especially where measurements are taken under automatic or computer control.

It is always useful to know how a digital multimeter works in order to be able to make the best use of it and the most accurate measurements. However it should be remembered that different multimeters from different manufacturers may work in different ways. It is therefore always helpful to consult the manufacturer›s instructions to understand how a particular digital multimeter works.

Precaution to be taken in handling digital Multi-meter,. Frequently occurring problems in Digital multimeters and the remedial measures

When using a digital multimeter, there are a number of initial steps and precautions that should be observed. These guidelines are always best to follow, and if incorporated into a workflow they will help to make measurements more accurate and prevent damage to the instrument or reduce safety risks for the user.

- ***Check battery regularly:*** DMMs require a power for their operation. For portable instruments, this is provided by a battery. Regular checks of the battery state are very advisable to ensure that there is sufficient voltage to adequately power the DMM. Also if batteries are left in situ for long periods they can leak and dame the contacts in the instrument as the contents of the battery are corrosive. To achieve this a sticker such as one indicating the data the battery is due for replacement or a calibration due sticker could be used.

- ***Return meter to high voltage setting after use:*** To prevent the possibility of accidental damage by the meter being set to a current or low voltage range, it is always wise to leave the meter set to read a high voltage, even if there is an Off button. It is too easy to automatically connect a digital multimeter to the circuit without thinking about the range This can lead to damage of both the equipment under test and the meter if it is set to a current range when voltage is to be measured, for example. Although some meters are auto-ranging for voltage, current etc, others are not, and therefore it is always wise to leave the meter set to the highest voltage range possible.
- ***Ensure probes are in good condition:*** Poor probes may not only result in poor readings, because it is not possible to connect to any test points properly, but also there can be the risk of injury if they are cracked and broken leaving exposed conductive areas when making a high voltage measurement.

Practical Exercise-Cleaning the switch contacts using Switch Cleaning Solution

What is a Switch Cleaning Solution- A solvent / lubricant blend designed to remove surface contamination and oxidation from all metallic contact surfaces. Reduces electrical resistance across the switch or connector by cleaning the surface. WD-40 is most popular Switch Cleaning Solution.

Key Properties: Cleans and lubricates switches, connectors and slip rings. Removes dirt and protects from further contamination. Reduces contact resistance. Commonly known as switch cleaner.

It is like a spray. Just spray over the rusty areas or greasy areas and leave it for 5 minutes. After that use the surface for any kind of Voltage measurement or soldering as required.

Practical Exercise: How to test a Fuse with a multimeter

1. Fuses are really just wires that are designed not to last, but their purpose is to prevent damage to more valuable electrical equipment or prevent fires (especially in homes) caused by power surges. If too much power runs through the fuse, it will "burn out," quite literally, and break the circuit. There are number of varieties of fuse, but their differences are primarily in appearance.
2. Take the FUSE out of the Circuit.
3. In the Analog or Digital Multimeter keep the Mode in Resistance Mode.
4. Keep the scale in the Lowest range.
5. Touch the probe of the meter on both ends of the Fuse.
6. The meter should read close to ZERO if the FUSE is OK .
7. The meter will read OPEN or OVERRANGE if the FUSE is bad.

Alternate Method with Continuity Testing

1. Keep the Meter in Continuity Testing Mode.
2. Touch the probe of the meter on both sides of the FUSE.
3. Sounding buzzer would indicate FUSE is good otherwise FUSE is bad.

Practical Exercise-Replace the FUSE with a matching rating

1. Check the Voltage rating of the Fuse.
2. Check the current rating of the Fuse.
3. Check the type of the Fuse- Whether it is Fast Blow or Slow Blow
4. Replace the Fuse with matching Voltage and Current Rating as well as with matching type.

Slow Blow Type Fuse will have spring type wiring inside and acts slowly.	**Fast Blow Fuse will have a single wire and it is acts very fast**

CHAPTER

3

Basic Electronics Components

Electronic components are of two types:

1. **Active Components:** Those devices or components which required external source to their operation is called Active Components.

   F*or Example:* Diode, Transistors, SCR etc...

   Explanation and Example: As we know that Diode is an Active Components. So it is required an External Source to its operation. Because, If we connect a Diode in a Circuit and then connect this circuit to the Supply voltage., then Diode will not conduct the current Until the supply voltage reach to 0.3(In case of Germanium) or 0.7V(In case of Silicon).

2. **Passive Components:** Those devices or components which do not required external source to their operation is called Passive Components.

   For Example: Resistor, Capacitor, Inductor etc

   Explanation and Example: Passive Components do not require external source to their operation. Like a Diode, Resistor does not require 0.3 0r 0.7 V. I.e., when we connect a resistor to the supply voltage, it starts work automatically without using a specific voltage. If you understood the above statement about active Components, then you will easily get this example.

Resistor

Definition, types of resistors, their construction & specific use, color-coding, power rating.

Definition of Resistors: These are passive electronic components which resists the flow of current is a circuit thereby regulating the flow of current in the circuit.

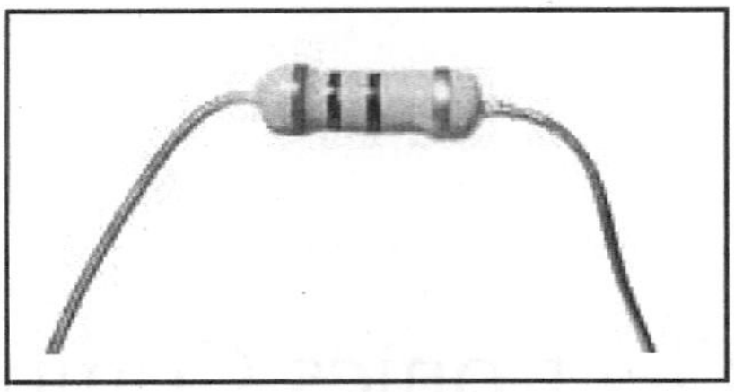

Figure 3.1: A typical resistor

Standard Resistor Symbols- The following are the standard Resistor Symbols

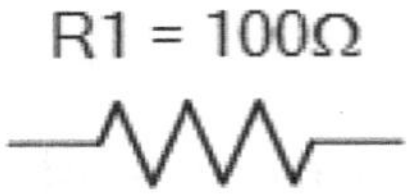

or

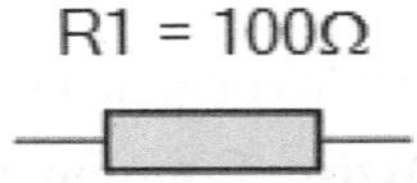

Zig- Zag type line symbol **Rectangular Box Symbol for Resistor**

Resistors can be Of Fixed Value or can have Variable Values called Potentiometer.

Fixed Resistors can be classified into four broad groups;

- ***Carbon Composition Resistor:*** Made of carbon dust or graphite paste, low wattage values
- ***Film Type Resistor:*** Made from conductive metal oxide paste, very low wattage values
- ***Wire-wound Resistor:*** Metallic bodies for heat sink mounting, very high wattage ratings
- ***Semiconductor Resistor:*** High frequency/precision surface mount thin film technology.

Practical Exercise: How to identify different types of resistors

Composition Types of Resistor

Carbon Resistors are the most common type of Composition Resistors. Carbon resistors are a cheap general purpose resistor used in electrical and electronic circuits. Their resistive element is manufactured from a mixture of finely ground carbon dust or graphite (similar to pencil lead) and a non-conducting ceramic (clay) powder to bind it all together.

Carbon Resistor

The ratio of carbon dust to ceramic (conductor to insulator) determines the overall resistive value of the mixture and the higher the ratio of carbon, the lower the overall resistance. The mixture is moulded into a cylindrical shape with metal wires or leads are attached to each end to provide the electrical connection as shown, before being coated with an outer insulating material and colour coded markings to denote its resistive value.

Carbon Resistor

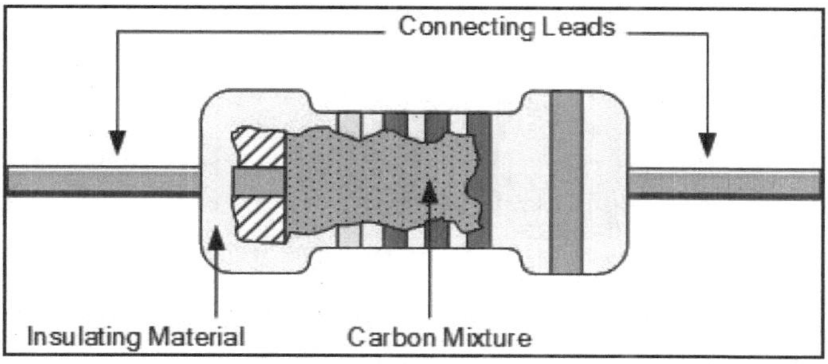

The **Carbon Composite Resistor** is a low to medium type power resistor which has a low inductance making them ideal for high frequency applications but they can also suffer from noise and stability when hot. Carbon composite resistors are generally prefixed with a "CR" notation (eg, CR10kΩ) and are available in E6 (± 20% tolerance (accuracy)), E12 (± 10% tolerance) and E24 (± 5% tolerance) packages with power ratings from 0.125 or 1/4 of a Watt up to 5 Watts.

Carbon composite resistor types are very cheap to make and are therefore commonly used in electrical circuits. However, due to their manufacturing process carbon type resistors have very large tolerances so for more precision and high value resistances, **film type resistors** are used instead.

Film Type Resistors

The generic term "Film Resistor" consist of *Metal Film*, *Carbon Film* and *Metal Oxide Film* resistor types, which are generally made by depositing pure metals, such as nickel, or an oxide film, such as tin-oxide, onto an insulating ceramic rod or substrate.

Film Resistor

The resistive value of the resistor is controlled by increasing the desired thickness of the deposited film giving them the names of either "thick-film resistors" or "thin-film resistors".

Once deposited, a laser is used to cut a high precision spiral helix groove type pattern into this film. The cutting of the film has the effect of increasing the conductive or resistive path, a bit like taking a long length of straight wire and forming it into a coil.

This method of manufacture allows for much closer tolerance resistors (1% or less) as compared to the simpler carbon composition types. The tolerance of a resistor is the difference between the preferred value (i.e, 100 ohms) and its actual manufactured value i.e, 103.6 ohms, and is expressed as a percentage, for example 5%, 10% etc, and in our example the actual tolerance is 3.6%. Film type resistors also achieve a much higher maximum ohmic value compared to other types and values in excess of 10MΩ (10 Million Ω's) are available.

Film Resistor

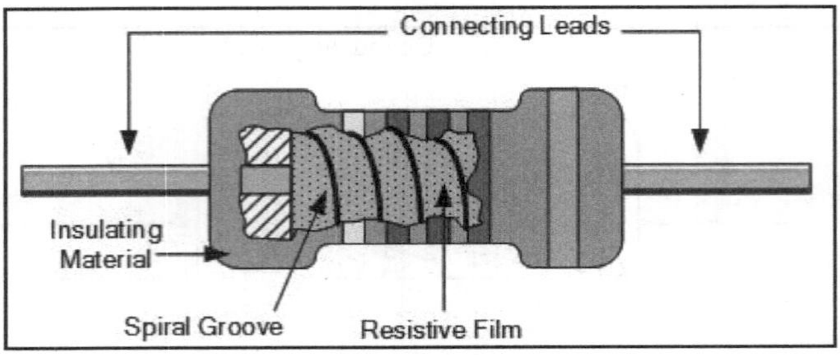

Film Resistor

Metal Film Resistors have much better temperature stability than their carbon equivalents, lower noise and are generally better for high frequency or radio frequency applications. **Metal Oxide Resistors** have better high surge current capability with a much higher temperature rating than the equivalent metal film resistors.

Another type of film resistor commonly known as a Thick Film Resistor is manufactured by depositing a much thicker conductive paste of CERamic and METal, called Cermet, onto an alumina ceramic substrate. Cermet resistors have similar properties to metal film resistors and are generally used for making small surface mount chip type resistors, multi-resistor networks in one package for pcb's and high frequency resistors. They have good temperature stability, low noise, and good voltage ratings but low surge current properties.

Metal Film Resistors are prefixed with a "MFR" notation (eg, MFR100kΩ) and a CF for Carbon Film types. Metal film resistors are available in E24 (±5% & ±2% tolerances), E96 (±1% tolerance) and E192(±0.5%, ±0.25% & ±0.1% tolerances) packages with power ratings of 0.05 (1/20th) of a Watt up to 1/2 Watt. Generally speaking Film resistors are precision low power components.

Wire wound Types of Resistor

Another type of resistor, called a **Wire wound Resistor**, is made by winding a thin metal alloy wire (Nichrome) or similar wire onto an insulating ceramic former in the form of a spiral helix similar to the film resistor above.

Wirewound Resistor

These types of resistor are generally only available in very low ohmic high precision values (from 0.01 to 100kΩ) due to the gauge of the wire and number of turns possible on the former making them ideal for use in measuring circuits and Whetstone bridge type applications.

They are also able to handle much higher electrical currents than other resistors of the same ohmic value with power ratings in excess of 300 Watts. These high power resistors are moulded or pressed into an aluminium heat sink body with fins attached to increase their overall surface area to promote heat loss and cooling.

These special types of resistor are called "Chassis Mounted Resistors" because they are designed to be physically mounted onto heatsinks or metal plates to further dissipate the generated heat. The mounting of the resistor onto a heatsink increases their current carrying capabilities even further.

Another type of wirewound resistor is the Power Wirewound Resistor. These are high temperature, high power non-inductive resistor types generally coated with a vitreous or glass epoxy enamel for use in resistance banks or DC motor/servo control and dynamic braking applications. They can even be used as low wattage space or cabinet heaters.

The non-inductive resistance wire is wound around a ceramic or porcelain tube covered with mica to prevent the alloy wires from moving when hot. Wirewound resistors are available in a variety of resistance and power ratings with one main use of power wirewound resistor is in the electrical heating elements of an electric fire which converts the electrical current flowing through it into heat with each element dissipating up to 1000 Watts, (1kW) of energy.

Because the wire of standard wire wound resistors is wound into a coil inside the resistors body, it acts like an inductor causing them to have inductance as well as resistance. This affects the way the resistor behaves in AC circuits by producing a phase shift at high frequencies especially in the larger size resistors. The length of the actual resistance path in the resistor and the leads contributes inductance in series with the "apparent" DC resistance resulting in an overall impedance path of ZOhms.

Impedance (Z) is the combined effect of resistance (R) and inductance (X), measured in ohms and for a series AC circuit is given as, Z 2 = R 2 + X 2.

When used in AC circuits this inductance value changes with frequency (inductive reactance,XL = $2\pi f$L) and therefore, the overall value of the resistor changes. Inductive reactance increases with frequency but is zero at DC (zero frequency). Then, wirewound resistors must not be designed or used in AC or amplifier type circuits where the frequency across the resistor changes. However, special non-inductive wirewound resistors are also available.

Wirewound Resistor

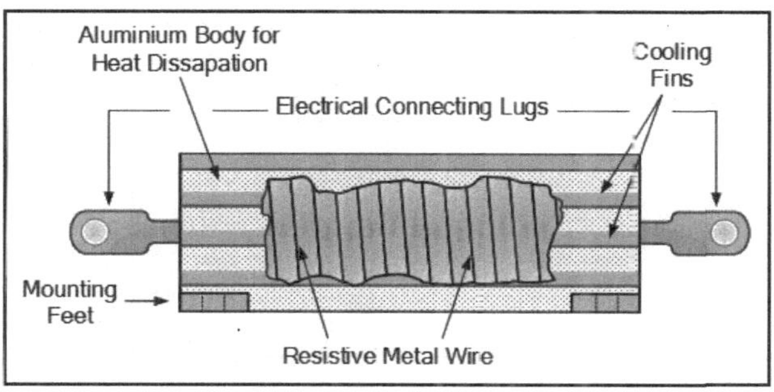

Wirewound resistor construction

Wirewound resistor types are prefixed with a "WH" or "W" notation (eg WH10Ω) and are available in the WH aluminium clad package (±1%, ±2%, ±5% & ±10% tolerance) or the W vitreous enamelled package (±1%, ±2% & ±5% tolerance) with power ratings from 1W to 300W or more.

The Colour Codes in a Resistor

There are colour bands on the resistances which is used to denote the value of Resistors. This way small space on the resistance is utilised to write the value of resistance on the Resistors. The colours are given the values as below:

Black- 0, Brown-1, Red-2, Orange-3, Yellow-4, Green- 5, Blue-6, Violet-7, Grey-8, White-9,

Gold and Silver are utilised to denote the tolerance values of 5% and 10% respectively.

How to use the code:

We have a FOUR Band scheme which is most common. First and second Bands are for the two digits and third band is for Multiplier and the fourth band is for tolerance.

Multiplier is $10^{\text{Color Code.}}$. For example if the third band is RED. The Multiplier will be $10^2 = 100$.

Thus the Multipliers for Black=10^0=1, For Brown=10^1=10 and so on

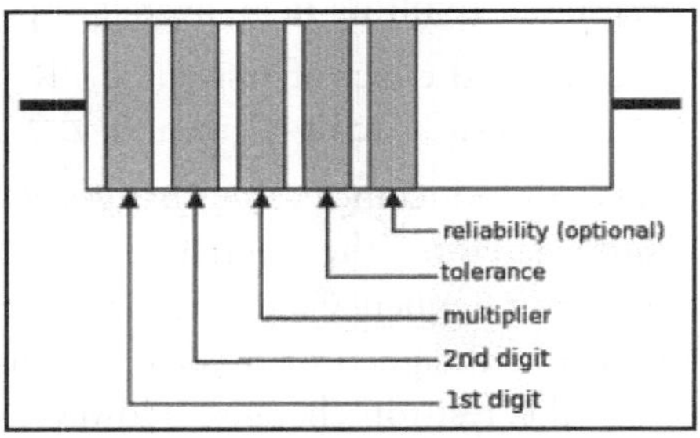

Practical Exercise: Measuring Resistance Values using colour codes and verify the readings by measuring in a Digital Multimeter

Example: Take the example of a resistor given below:

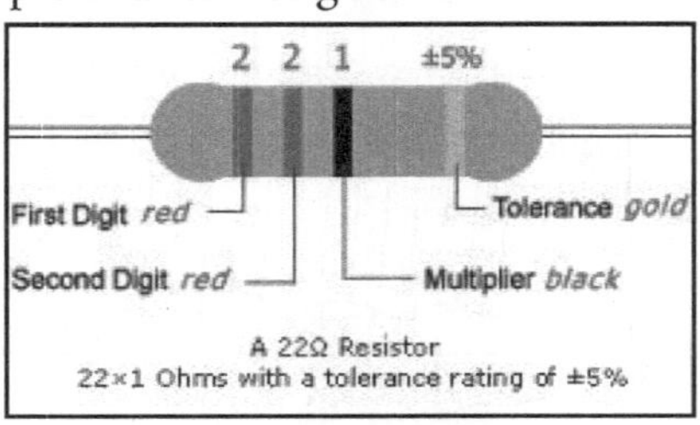

Resistor Band Codes

First Digit= 2 (code for RED), Second Digit= 2(Code for RED), Third band is Multiplier= 10^0=1 and the Fourth band is Gold indicating a tolerance of 5%.

Thus the value of this Resistor will be : 2 2 x 1= 22 Ohms having tolerance of 5%.

Verify the readings by measuring the resistance as described in Multimeter section.

Series and Parallel connections of Resistors

Series Resistor Connections

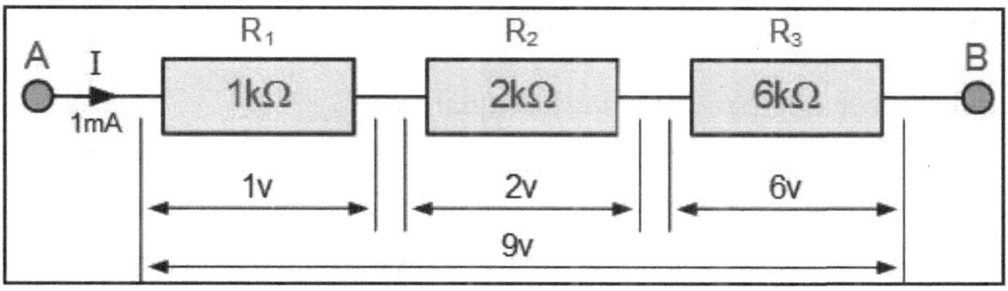

Series Resistor Connections

- Resistors that are daisy chained together in a single line are said to be connected in **SERIES**. One end of the Resistor is connected to the other end of the other and so on.
- Series connected resistors have a common **Current** flowing through them.

  $I_{total} = I1 = I2 = I3 .... etc$
- The total circuit resistance of series resistors is equal to:

  $R_{total} = R_1 + R_2 + R_3 + ... R_n$ etc.
- Total circuit voltage is equal to the sum of all the individual voltage drops.

  $V_{total} = V1 + V2 + V3 .... etc$
- The total resistance of a series connected circuit will always be greater than the highest value resistor.

Parallel Resistor Connection

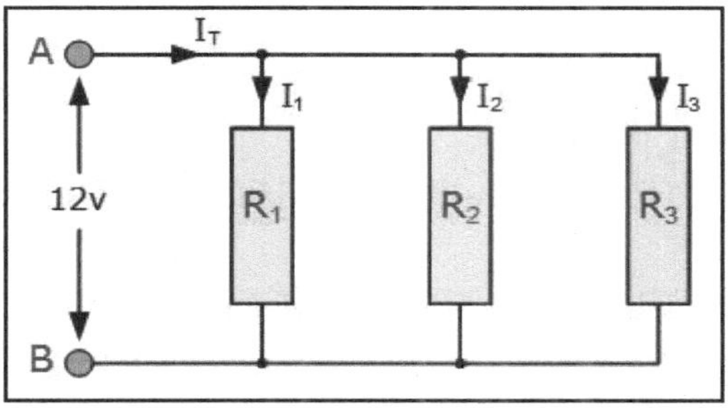

Resistance connected in parallel

Resistors in Parallel

Resistors are said to be connected together in "Parallel" when both of their terminals are respectively connected to each terminal of the other resistor or resistors. Unlike the previous series resistor circuit, in a parallel resistor network the circuit current can take more than one path as there are multiple paths for the current. Then parallel circuits are classed as current dividers.

- Resistors that have both of their respective terminals connected to each terminal of another resistor or resistors are said to be connected in **PARALLEL**.
- Parallel resistors have a common **Voltage** across them.

  $V_{S = V1 = V2 = V3 \ldots. etc}$
- Total resistance of a parallel circuit is equal to:

  $$\frac{1}{R_T} = \frac{1}{R_1} + \frac{1}{R_2} + \frac{1}{R_3} \ldots. + \frac{1}{R_n} \text{ etc.}$$
- Total circuit current flow is equal to the sum of all the individual branch currents added together.

  $I_{total = I1 + I2 + I3 \ldots. etc}$
- The total resistance of a parallel circuit will always be less than the value of the smallest resistor.

Resistor Power Rating

- The larger the power rating, the greater the physical size of the resistor to dissipate the heat.
- All resistors have a maximum power rating and if exceeded will result in the resistor overheating and becoming damaged.
- Standard resistor power rating sizes are 1/8 W, 1/4 W, 1/2 W, 1 W, and 2 W.
- Low ohmic value power resistors are generally used for current sensing or power supply applications.
- The power rating of resistors can be calculated using the formula:

  $$\text{Power (P)} = V x I = I^2 R = \frac{V^2}{R}$$
- In AC Circuits the voltage and current flowing in a pure resistor are always "*in-phase*" producing 0° phase shift..
- When used in AC Circuits the AC impedance of a resistor is equal to its DC Resistance.
- The AC circuit impedance for resistors is given the symbol Z.

Practical Exercise: Identifying Power Rating of a resistor by using size

As the dissipated resistor power rating is linked to their physical size, a 1/4 (0.250) W resistor is physically smaller than a 1W resistor, and resistors that are of the same ohmic value are also available in different power or wattage ratings. Carbon resistors, for example, are commonly made in wattage ratings of 1/8 (0.125)W, 1/4 (0.250)W, 1/2 (0.5)W, 1W, and 2 Watts.

Generally speaking the larger their physical size the higher its wattage rating.

Wirewound power resistors come in a variety of designs and types, from the standard smaller heatsink mounted aluminium body 25W types as we have seen previously, to the larger tubular 1000W ceramic or porcelain power resistors used for heating elements.

CHAPTER 4

Inductors

TYPES OF INDUCTORS, SPECIFICATIONS AND APPLICATIONS

The Inductor

The Inductor is a Coil of WIRES. This device stores Energy in the Form of Magnetic Energy . The Energy exists till there is current Flowing through it. When the current is not the Energy will not be there i.e it will no more a Magnet. Inductor opposes the change in Electric current.

[When an electrical current flows through a wire conductor, a magnetic flux is developed around the conductor producing a relationship between the direction of this magnetic flux which is circulating around the conductor and the direction of the current flowing through the same conductor. This well known relationship between current and magnetic flux direction is called, "Fleming's Left Hand Rule". This phenomenon is known as Electromagnetism.]

Practical Experiment: Make an Electromagnet with a Coil

Make a circuit as below:

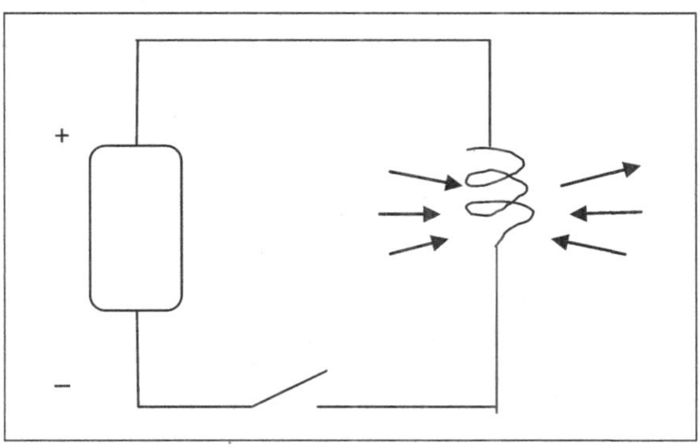

Figure 4.1: Experiment of an Electromagnet

Connect a battery with a switch and an Coil.

Take some Pins.

Keep the switch in OFF position and bring the pins near the coil. Nothing happens.

Keep the switch in ON position and bring the Pins near the coil.

You will observe that the coil has become a magnet as Electric current flows through it and it attracts the Pins.

But there is also another important property relating to a wound coil that also exists, which is that a secondary voltage is induced into the same coil by the movement of the magnetic flux as it opposes or resists any changes in the electrical current flowing it.

Thus point to remember is the following:

1. When Electric current flows through the coil it develops a magnetic field around it.
2. The magnetic field keeps on expanding.
3. The changing magnetic field then generates a Voltage and a current which flows opposite to the existing Voltage. Thus the inductor affects the rate of rise and fall of current .
4. This happens till the magnetic field is stabilised and then there is no opposing current develops. The current develops because of changing magnetic field.
5. The Coil or Inductor then acts like a Short circuit act as Zero resistance and will pass all the Current.
6. The inductor DELAYS the current build up and Stores the Energy in the form of Magnetic Field.

Experiment: To Study the Current Flow Delay Action of the Inductor

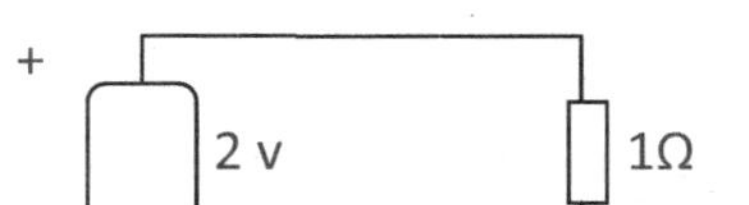	I=V/R= 2 V/1Ω= 2 A. If we measure the current it will be 2 A immediately when the switch is closed and it will be 0 immediately when the switch is Closed. If we replace the Resistor with a Bulb the Bulb will immediately GLOW to full brightness on switch ON and will be OFF immediately when the Switch is OFF.

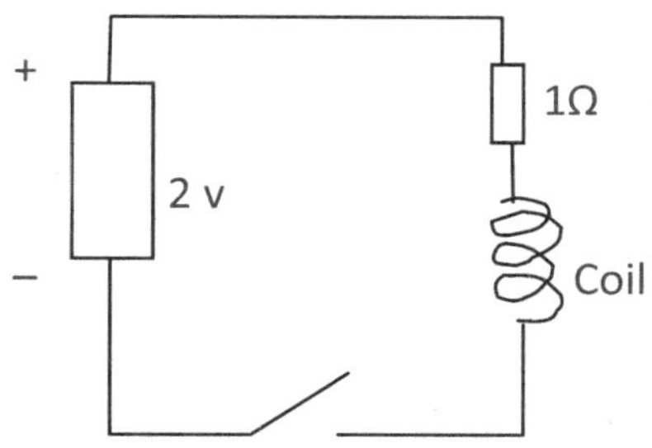

Now a Coil is introduced in the Circuit. If we measure the current in the Circuit it will not RISE immediately to 2 A. Rather it will take some time to reach to 2 A. Similarly when we put the Switch OFF the current will still be there and will become 0 after some DELAY. Replace the Resistor with a BULB. The bulb is very dim at the time of Switch on and GLOWS to full brightness only after some delays . Similarly on switch OFF it becomes DIM and becomes OFF only after DELAY.

Compare the action to accelerator of a car. The car accelerates and gains speed only after delayed time then the speed is Stabilised.

In its most basic form, an Inductor is nothing more than a coil of wire wound around a central core. For most coils the current, (i) flowing through the coil produces a magnetic flux, (NΦ) around it that is proportional to this flow of electrical current.

Figure 4.2 A Typical Inductor

The Inductor, also called a choke, is another passive type electrical component which is just a coil of wire that is designed to take advantage of this relationship by inducing a magnetic field in itself or in the core as a result of the current passing through the coil. This results in a much stronger magnetic field than one that would be produced by a simple coil of wire.

Inductors are formed with wire tightly wrapped around a solid central core which can be either a straight cylindrical rod or a continuous loop or ring to concentrate their magnetic flux.

The schematic symbol for a inductor is that of a coil of wire so therefore, a coil of wire can also be called an **Inductor**.

Inductors are usually are categorised according to the type of inner core they are wound around, for example, hollow core (free air), solid iron core or soft ferrite core with the different core types being distinguished by adding continuous or dotted parallel lines next to the wire coil as shown below.

Inductor Symbols

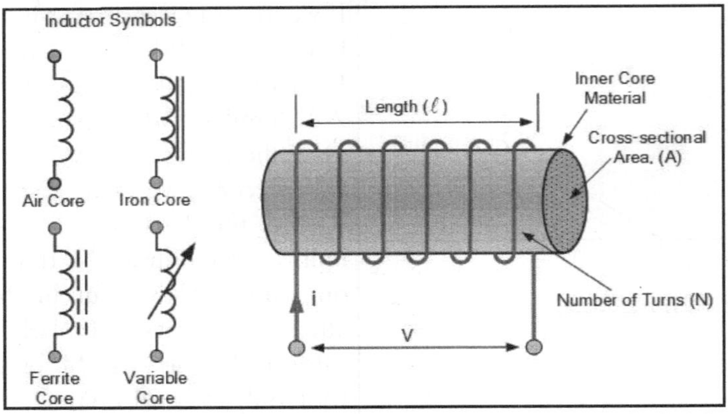

The current, *i* that flows through an inductor produces a magnetic flux that is proportional to it. But unlike a **Capacitor** which oppose a change of voltage across their plates, an inductor opposes the rate of change of current flowing through it due to the build up of self-induced energy within its magnetic field.

In other words, inductors resist or oppose changes of current but will easily pass a steady state DC current. This ability of an inductor to resist changes in current and which also relates current, i with its magnetic flux linkage, NΦ as a constant of proportionality is called Inductance which is given the symbol L with units of Henry, (H) after Joseph Henry.

Because the Henry is a relatively large unit of inductance in its own right, for the smaller inductors sub-units of the Henry are used to denote its value. For example:

Inductance Prefixes

Prefix	Symbol	Multiplier	Power of Ten
milli	m	1/1,000	10^{-3}
micro	µ	1/1,000,000	10^{-6}
nano	N	1/1,000,000,000	10^{-9}

So to display the sub-units of the Henry we would use as an example:

- 1mH = 1 milli-Henry – which is equal to one thousandths (1/1000) of an Henry.
- 100uH = 100 micro-Henries – which is equal to 100 millionth's (1/1,000,000) of a Henry.

The capacity of the inductors is controlled by the following:

- The number of turns of Coils.
- The materials that the Coils are wrapped around called CORE.

- The cross sectional area of the coil. The larger the cross section the longer the Inductance.

Practical Exercise: Dismantle and identify the different parts of a relay

There are four main parts in a relay. They are

1. Electromagnet
2. Movable Armature
3. Switch point contacts
4. Spring

The figures given below show the dismantled parts of a simple relay.

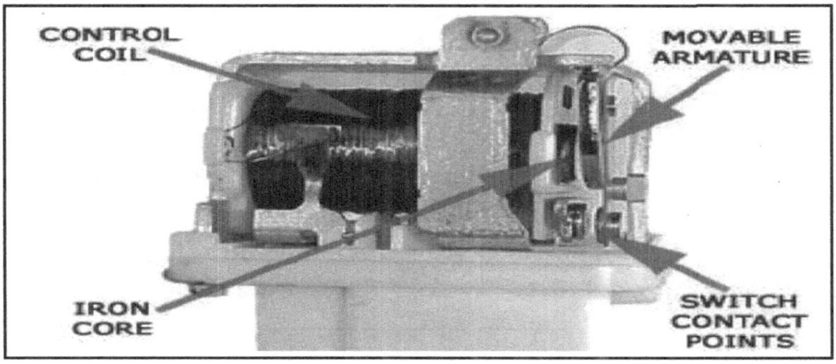

Relay Construction

It is an electro-magnetic relay with a wire coil, surrounded by an iron core. When electricity flows through the coil it produces magnetic flux and attracts the armature and close the contact to the switch. These parts are safely held with the help of a spring.

Practical –Identify Different Types of Inductors

A diverse electronic component used in a wide range of applications requires various types of inductors. These are of different shapes, sizes including the wire wound and multilayer inductors. Different types of inductors include high-frequency inductors, power supply line inductors or power inductors and inductors for general circuits. Differentiation of the inductors is based on the type of winding as well as the core used

Air Core Inductors

In air core inductors the Core is missing. These type of inductors are large in size as they can't offer more inductance without a Core. These are used in Radio and TV Receivers and other High Frequency applications.

Figure 4.3: Air Core Inductors

Ferro Magnetic or Iron Core Inductors

The Iron Core inductors have Iron as their Core. These are not suitable for High Frequency applications as they have Eddy Current and Hysteresis losses. The Transformer is an example of Iron Core Inductors.

Figure 4.4: Iron Core Inductors

Ferrite Core Inductors

Ferrite Core Inductors have Ferrite (a metal oxide ceramic based around a mixture of Ferric Oxide Fe2O3). These have small sizes and are good for High Frequency applications due to low core losses.

Figure 4.5: Ferrite Core Inductors

Toroidal Core Inductors

Figure 4.6: Toroidal Core Inductor

In this type of inductors the Coil is wounded on a circular Core of Iron or Ferrite core. A special winding machine is required. This offers high inductance in a small space because of very small losses.

Bobbin based inductors

Figure 4.7: Bobbin based inductors

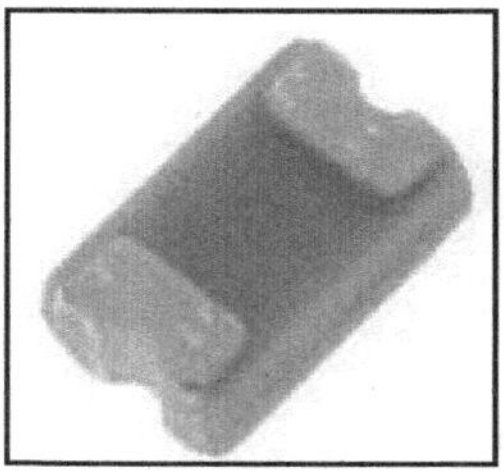

Figure 4.8: Multilayerd Inductor

The coil is wounded on the Bobbin. This is used in Switched Mode Power Supply and Power converters.

Multilayered inductors

A multilayer inductor contains a first conductive coil pattern and a second conductive coil pattern which are arranged in two layers in the upper part of a multi-layered body and are connected electrically in a consecutive manner in series to 2 more conductive coil patterns disposed in the lower part of the multi-layered body and therefore forming a spiral coil. These are mainly used in Mobile devices and also in Switched Mode Power Supplies and small DC-DC Converters.

Thin Film Inductors

Figure 4.9: Thin Film Inductors

These are completely different from the conventional chip-type inductors wound with copper wire. In this type, tiny inductors are formed using thin-film processing to create the chip inductor for high-frequency applications, which ranges from about nano Henry.

What is Reactance: As the resistor limits the current so does the REACTANCE in an inductor limits the flow of current. The formula for reactance is as below:

$X_L = 2\pi fL$ Ohms

Where X_L is Inductance in Ohms, f is frequency in Hertz and L is inductance in Henry.

For DC Currents: The Value of f will be Zero. Thus the Reactance offered is 0. Remember that for DC the Inductor acts like a SHORT CIRCUIT just like a WIRE.

For AC Circuits: The value of f will increase and as the frequency increases the REACTANCE increases. Remember thus that for High Frequency circuits the Inductor acts like an OPEN CIRCUIT.

Applications of Inductors

1. Application of CHOKE in Tubelights:

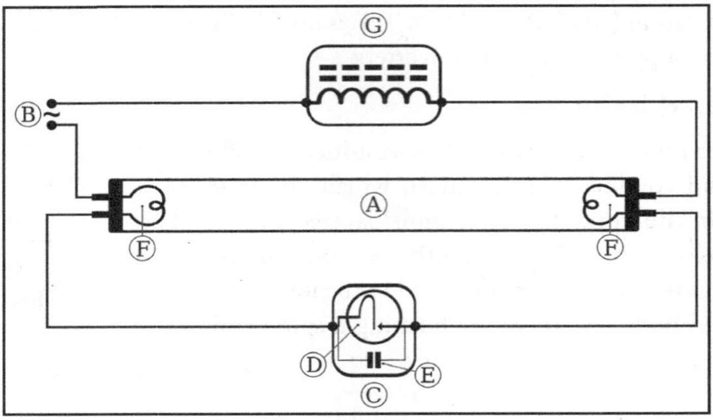

Figure 4.10: A typical Tubelight Circuit

A: Fluorescent tube, B: Power (+220 volts), C: Starter, D: Switch (bi-metallic thermostat), E: Capacitor, F: Filaments, G: Ballast

CHOKE is nothing but an Inductor and Starter consists of a Bimetallic Thermostat Switch and a Capacitor.

In a tube light, Ionization should take place inside tube to get Ions produces which induces light, To get Ionization needs high voltage at starting , Once it gets ionized, after that no need of High voltage, It mean at starting we require high voltage. Tube light will have Choke (which is an Inductor) and starter contains capacitor in parallel with a Bimetallic switch.

When you switch on starter bulb is short circuited and heating current flows through series circuit. After some time (few second) gas bulb is heated enough and bi-metal electrode break the series circuit. Sudden reduce in circuit current causes induction of high voltage on inductor and so on the opposite electrodes of neon tube. Neon tube starts to glow giving the light. At the same time the starter bulb continue to glow and keep hot preventing bi-metal electrode to close again. Start of neon tube may be clean or through few attempts (flickering). Inductor limits the neon tube voltage when it glows to abt. 50V.

The capacitor is for EMI suppression(which is of high frequency) and will reduce the humming sound. We will learn in the Capacitor section that Capacitor acts as short circuit for High Frequencies. This is typically a fairly-small value - anywhere between 1n to 100n, depending upon who made your particular starter.

2. ***APPLICATION AS Filters:*** Inductors are used extensively with capacitors and resistors to create filters for analog circuits and in signal processing. Alone, an inductor functions as a low-pass filter, since the impedance of an inductor increases as the frequency of a signal increases. When combined with a capacitor, whose impedance decreases as the frequency of a signal increase, a notched filter can be made that only allows a certain frequency range to pass through. By

combining capacitors, inductors, and resistors in a number of ways advanced filter topologies can be created for any number of applications. Filters are used in most electronics, although capacitors are often used rather than inductors when possible since they are smaller and cheaper.

3. ***As sensors in Traffic light controllers:*** Inductors can be used to sense magnetic fields or the presence of magnetically permeable material from a distance. Inductive sensors are used at nearly every intersection with a traffic light to detect the amount of traffic and adjust the signal accordingly.
4. ***As Transformers:*** Combining inductors that have a shared magnetic path will form a transformer. The transformer is a fundamental component of national electrical grids and found in many power supplies as well to increase or decrease voltages to a desired level. Since magnetic fields are created by a change in current, the faster the current changes (increase in frequency) the more effective a transformer operates. Of course, as the frequency of the input increases, the impedance of the inductor begins to limit the effectiveness of a transformer. Practically inductance based transformers are limited to the 10s of kHz, usually lower. The benefit of a higher operating frequency is a smaller and lighter weight transformer can be used to deliver the same load.
6. ***Motors:*** Normally inductors are in a fixed position and not allowed to move to align themselves with any nearby magnetic field. Inductive motor leverage the magnetic force applied to inductors to turn electrical energy in to mechanical energy. Inductive motors are designed so that a rotating magnetic field is created in time with an AC input. Since the speed of rotation is controlled by the input frequency, induction motors are often used in fixed speed applications that can be powered directly from 50/60hz mains power. The biggest advantage of inductive motors over other designs is that no electrical contact is required between the rotor and the motor which makes inductive motors very robust and reliable.
7. ***Energy Storage:*** Like capacitors, inductors can be used for energy storage. Unlike capacitors, inductors have a severe limitation on how long they can store energy since the energy is stored in a magnetic field which collapses quickly once power is removed. The main use for inductors as energy storage is in switch-mode power supplies, like the power supply in a PC. In the simpler, non-isolated switch-mode power supplies, a single inductor is used in place of transformer and energy storage component. In these circuits, the ratio of the time the inductor is powered to the time it is unpowered determines the input to output voltage ratio.

CHAPTER

5

Capacitors

TYPES OF CAPACITORS, SPECIFICATIONS AND APPLICATIONS

Introduction to Capacitors

Just like the Resistor, the Capacitor, sometimes referred to as a Condenser, is a simple passive device that is used to "store electricity". The capacitor is a component which has the ability or "capacity" to store energy in the form of an electrical charge producing a potential difference (Static Voltage) across its plates, much like a small rechargeable battery.

There are many different kinds of capacitors available from very small capacitor beads used in resonance circuits to large power factor correction capacitors, but they all do the same thing, they store charge.

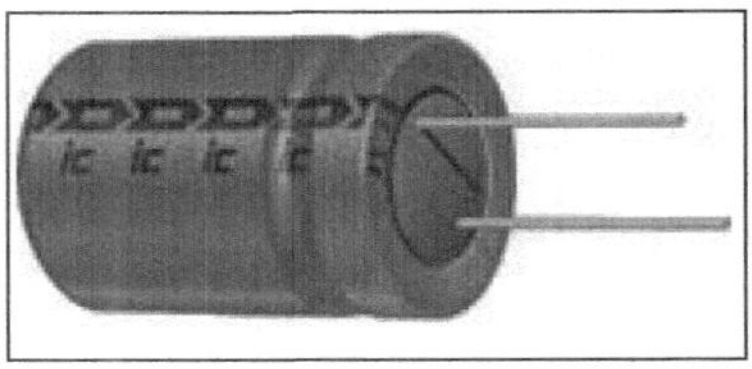

A Typical Capacitor

In its basic form, a Capacitor consists of two or more parallel conductive (metal) plates which are not connected or touching each other, but are electrically separated either by air or by some form of a good insulating material such as waxed paper, mica, ceramic, plastic or some form of a liquid gel as used in electrolytic capacitors. The insulating layer between a capacitors plates is commonly called the Dielectric.

Due to this insulating layer, DC current can not flow through the capacitor as it blocks it allowing instead a voltage to be present across the plates in the form of an electrical charge.

The conductive metal plates of a capacitor can be either square, circular or rectangular, or they can be of a cylindrical or spherical shape with the general shape, size and construction of a parallel plate capacitor depending on its application and voltage rating.

When used in a direct current or DC circuit, a capacitor charges up to its supply voltage but blocks the flow of current through it because the dielectric of a capacitor is non-conductive and basically an insulator. However, when a capacitor is connected to an alternating current or AC circuit, the flow of the current appears to pass straight through the capacitor with little or no resistance.

There are two types of electrical charge, positive charge in the form of Protons and negative charge in the form of Electrons. When a DC voltage is placed across a capacitor, the positive (+ve) charge quickly accumulates on one plate while a corresponding negative (-ve) charge accumulates on the other plate. For every particle of +ve charge that arrives at one plate a charge of the same sign will depart from the -ve plate.

Then the plates remain charge neutral and a potential difference due to this charge is established between the two plates. Once the capacitor reaches its steady state condition an electrical current is unable to flow through the capacitor itself and around the circuit due to the insulating properties of the dielectric used to separate the plates.

The flow of electrons onto the plates is known as the capacitors **Charging Current** which continues to flow until the voltage across both plates (and hence the capacitor) is equal to the applied voltage Vc. At this point the capacitor is said to be "fully charged" with electrons. The strength or rate of this charging current is at its maximum value when the plates are fully discharged (initial condition) and slowly reduces in value to zero as the plates charge up to a potential difference across the capacitors plates equal to the source voltage.

The amount of potential difference present across the capacitor depends upon how much charge was deposited onto the plates by the work being done by the source voltage and also by how much capacitance the capacitor has and this is illustrated below.

Capacitor Construction

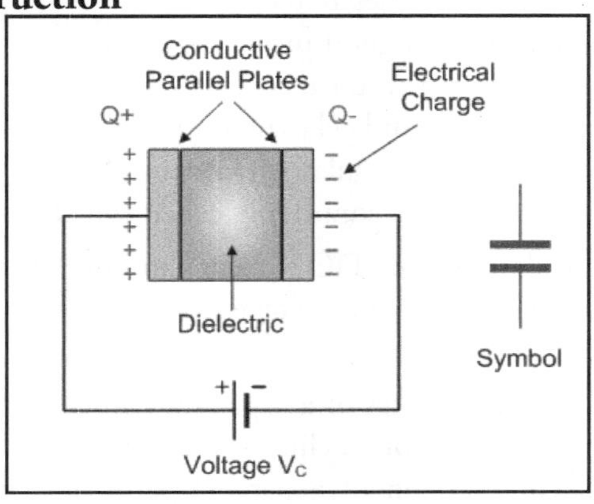

Figure 5.1: Construction of a Capacitor

The parallel plate capacitor is the simplest form of capacitor. It can be constructed using two metal or metallised foil plates at a distance parallel to each other, with its capacitance value in Farads, being fixed by the surface area of the conductive plates and the distance of separation between them. Altering any two of these values alters the the value of its capacitance and this forms the basis of operation of the variable capacitors.

Also, because capacitors store the energy of the electrons in the form of an electrical charge on the plates the larger the plates and/or smaller their separation the greater will be the charge that the capacitor holds for any given voltage across its plates. In other words, larger plates, smaller distance, more capacitance.

By applying a voltage to a capacitor and measuring the charge on the plates, the ratio of the charge Q to the voltage V will give the capacitance value of the capacitor and is therefore given as: $C = Q/V$ this equation can also be re-arranged to give the more familiar formula for the quantity of charge on the plates as: $Q = C \times V$

Although we have said that the charge is stored on the plates of a capacitor, it is more correct to say that the energy within the charge is stored in an "electrostatic field" between the two plates. When an electric current flows into the capacitor, charging it up, the electrostatic field becomes more stronger as it stores more energy. Likewise, as the current flows out of the capacitor, discharging it, the potential difference between the two plates decreases and the electrostatic field decreases as the energy moves out of the plates.

The property of a capacitor to store charge on its plates in the form of an electrostatic field is called the **Capacitance** of the capacitor. Not only that, but capacitance is also the property of a capacitor which resists the change of voltage across it.

The Capacitance of a Capacitor

Capacitance is the electrical property of a capacitor and is the measure of a capacitors ability to store an electrical charge onto its two plates with the unit of capacitance being the **Farad** (abbreviated to F) named after the British physicist Michael Faraday.

Capacitance is defined as being that a capacitor has the capacitance of One Farad when a charge of One Coulomb is stored on the plates by a voltage of One volt. Capacitance, C is always positive and has no negative units. However, the Farad is a very large unit of measurement to use on its own so sub-multiples of the Farad are generally used such as micro-farads, nano-farads and pico-farads, for example.

STANDARD UNITS OF CAPACITANCE

- Microfarad (μF) $1\mu F = 1/1{,}000{,}000 = 0.000001 = 10^{-6}$ F
- Nanofarad (nF) $1nF = 1/1{,}000{,}000{,}000 = 0.000000001 = 10^{-9}$ F
- Picofarad (pF) $1pF = 1/1{,}000{,}000{,}000{,}000 = 0.000000000001 = 10^{-12}$ F

Capacitance of a Parallel Plate Capacitor

The capacitance of a parallel plate capacitor is proportional to the area, A of the plates and inversely proportional to their distance or separation, d (i.e. the dielectric thickness) giving us a value for capacitance of C = k(A/d) where in a vacuum the value of the constant k is 8.84 x 10-12 F/m or 1/4.π.9 x 109, which is the permittivity of free space. Generally, the conductive plates of a capacitor are separated by air or some kind of insulating material or gel rather than the vacuum of free space.

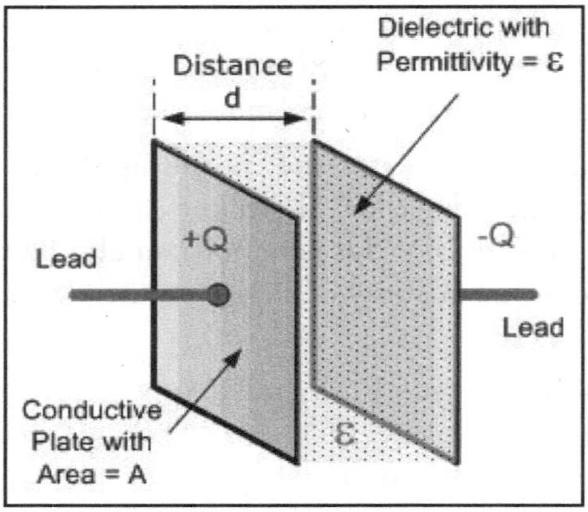

Figure 5.2: Construction of a parallel plate capacitor

The Dielectric of a Capacitor

As well as the overall size of the conductive plates and their distance or spacing apart from each other, another factor which affects the overall capacitance of the device is the type of dielectric material being used.

In other words the "Permittivity" (ε) of the dielectric. The conductive plates are generally made of a metal foil or a metal film but the dielectric material is an insulator.

The various insulating materials used as the dielectric in a capacitor differ in their ability to block or pass an electrical charge.

This dielectric material can be made from a number of insulating materials or combinations of these materials with the most common types used being: air, paper, polyester, polypropylene, Mylar, ceramic, glass, oil, or a variety of other materials.

The factor by which the dielectric material, or insulator, increases the capacitance of the capacitor compared to air is known as the Dielectric Constant, k and a dielectric material with a high dielectric constant is a better insulator than a dielectric material with a lower dielectric constant. Dielectric constant is a dimensionless quantity since it is relative to free space.

The actual permittivity or "complex permittivity" of the dielectric material between the plates is then the product of the permittivity of free space (ε_o) and the relative permittivity (ε_r) of the material being used as the dielectric and is given as:

One method used to increase the overall capacitance of a capacitor is to "interleave" more plates together within a single capacitor body. Instead of just one set of parallel plates, a capacitor can have many individual plates connected together thereby increasing the area, A of the plate. For example, a capacitor with 10 interleaved plates would produce 9 (10 – 1) mini capacitors with an overall capacitance nine times that of a single parallel plate.

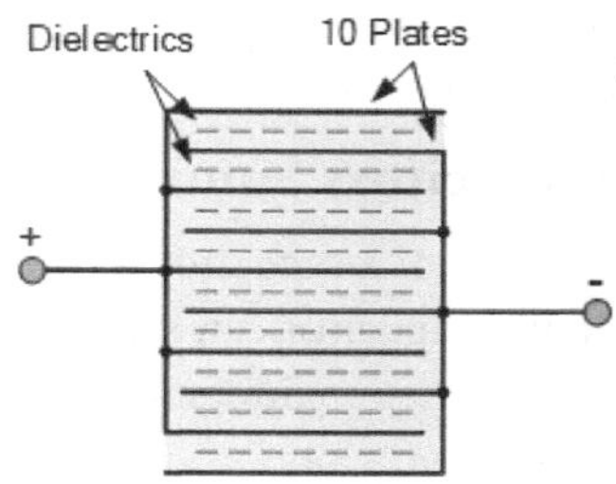

9 mini capacitors in one

Modern capacitors can be classified according to the characteristics and properties of their insulating dielectric:

- Low Loss, High Stability such as Mica, Low-K Ceramic, Polystyrene.
- Medium Loss, Medium Stability such as Paper, Plastic Film, High-K Ceramic.
- Polarized Capacitors such as Electrolytic's, Tantalum's.

Voltage Rating of a Capacitor

All capacitors have a maximum voltage rating and when selecting a capacitor consideration must be given to the amount of voltage to be applied across the capacitor. The maximum amount of voltage that can be applied to the capacitor without damage to its dielectric material is generally given in the data sheets as: WV, (working voltage) or as WV DC, (DC working voltage).

If the voltage applied across the capacitor becomes too great, the dielectric will break down (known as electrical breakdown) and arcing will occur between the capacitor plates resulting in a short-circuit. The working voltage of the capacitor depends on the type of dielectric material being used and its thickness.

The DC working voltage of a capacitor is just that, the maximum DC voltage and NOT the maximum AC voltage as a capacitor with a DC voltage rating of 100 volts DC cannot be safely subjected to an alternating voltage of 100 volts. Since an alternating voltage has an r.m.s. value of 100 volts but a peak value of over 141 volts!.Then a capacitor which is required to operate at 100 volts AC should have a working voltage of at least 200 volts. In practice, a capacitor should be selected so that its working voltage either DC or AC should be at least 50 percent greater than the highest effective voltage to be applied to it.

Another factor which affects the operation of a capacitor is Dielectric Leakage. Dielectric leakage occurs in a capacitor as the result of an unwanted leakage current which flows through the dielectric material.

Generally, it is assumed that the resistance of the dielectric is extremely high and a good insulator blocking the flow of DC current through the capacitor (as in a perfect capacitor) from one plate to the other.

However, if the dielectric material becomes damaged due excessive voltage or over temperature, the leakage current through the dielectric will become extremely high resulting in a rapid loss of charge on the plates and an overheating of the capacitor eventually resulting in premature failure of the capacitor. Then never use a capacitor in a circuit with higher voltages than the capacitor is rated for otherwise it may become hot and explode.

Capacitors Summary

The job of a capacitor is to store charge onto its plates. The amount of electrical charge that a capacitor can store on its plates is known as its **Capacitance** value and depends upon three main factors.

- ***Surface Area*** – the surface area, A of the two conductive plates which make up the capacitor, the larger the area the greater the capacitance.
- ***Distance*** – the distance, d between the two plates, the smaller the distance the greater the capacitance.
- ***Dielectric Material*** – the type of material which separates the two plates called the "dielectric", the higher the permittivity of the dielectric the greater the capacitance.

The dielectric of a capacitor is a non-conducting insulating material, such as waxed paper, glass, mica different plastics etc, and provides the following advantages:

- The dielectric constant is the property of the dielectric material and varies from one material to another increasing the capacitance by a factor of k.
- The dielectric provides mechanical support between the two plates allowing the plates to be closer together without touching.
- Permittivity of the dielectric increases the capacitance.
- The dielectric increases the maximum operating voltage compared to air.

Capacitors can be used to block DC current while passing audio signals, pulses, or alternating current, or other time varying wave forms. This ability to block DC currents enables capacitors to be used to smooth the output voltages of power supplies, to remove unwanted spikes from signals that would otherwise tend to cause damage or false triggering of semiconductors or digital components. Capacitors can also be used to adjust the frequency response of an audio circuit, or to couple together separate amplifier stages that must be protected from the transmission of DC current.

At DC a capacitor has infinite impedance (open -circuit), at very high frequencies a capacitor has zero impedance (short-circuit). All capacitors have a maximum working voltage rating, its WV DC so select a capacitor with a rating at least 50% more than the supply voltage.

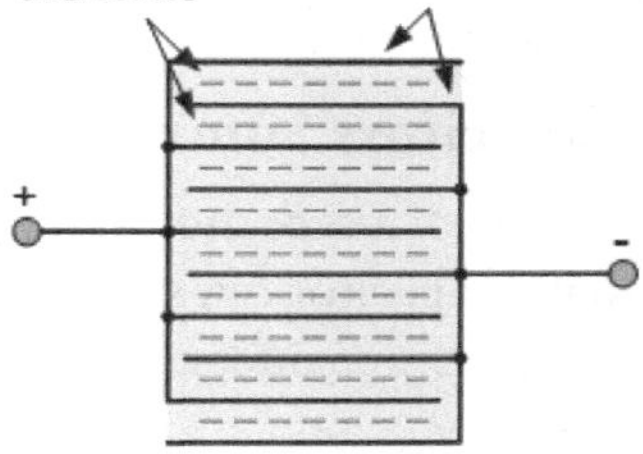

Practical Exercise: Identify different Types of capacitors

Variable Capacitor Symbols- Die electric capacitor

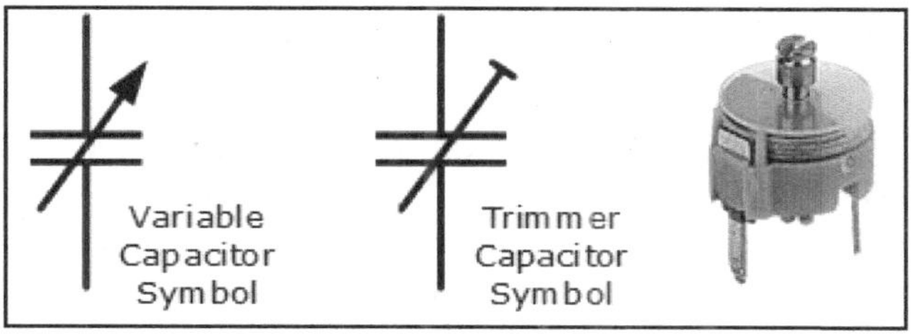

Continuously variable types AND preset type variable capacitors are available called Trimmers. These are generally small devices that can be adjusted or "pre-set" to a particular capacitance value with the aid of a small screwdriver and are available in very small capacitance's of 500pF or less and are non-polarized.

Film Capacitor

The film and foil types of capacitors are made from long thin strips of thin metal foil with the dielectric material sandwiched together which are wound into a tight roll and then sealed in paper or metal tubes.

These film types require a much thicker dielectric film to reduce the risk of tears or punctures in the film, and is therefore more suited to lower capacitance values and larger case sizes.

Metalised foil capacitors have the conductive film metalised sprayed directly onto each side of the dielectric which gives the capacitor self-healing properties and can therefore use much thinner dielectric films. This allows for higher capacitance values and smaller case sizes for a given capacitance. Film and foil capacitors are generally used for higher power and more precise applications.

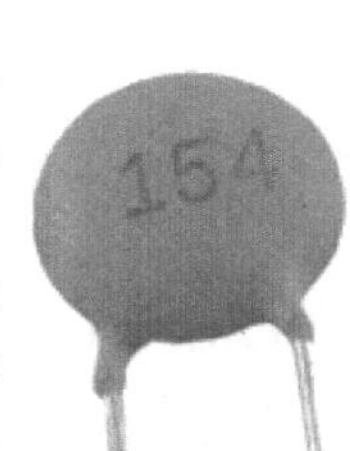

Ceramic Capacitor

Ceramic Capacitors or **Disc Capacitors** as they are generally called, are made by coating two sides of a small porcelain or ceramic disc with silver and are then stacked together to make a capacitor. For very low capacitance values a single ceramic disc of about 3-6mm is used. Ceramic capacitors have a high dielectric constant (High-K) and are available so that relatively high capacitance's can be obtained in a small physical size.

They exhibit large non-linear changes in capacitance against temperature and as a result are used as de-coupling or by-pass capacitors as they are also non-polarized devices. Ceramic capacitors have values ranging from a few picofarads to one or two microfarads, (μF) but their voltage ratings are generally quite low.

Ceramic types of capacitors generally have a 3-digit code printed

Electrolytic Capacitors are generally used when very large capacitance values are required. Here instead of using a very thin metallic film layer for one of the electrodes, a semi-liquid electrolyte solution in the form of a jelly or paste is used which serves as the second electrode (usually the cathode).

Electrolytic Capacitor

The majority of electrolytic types of capacitors are **Polarised**, that is the DC voltage applied to the capacitor terminals must be of the correct polarity, i.e. positive to the positive terminal and negative to the negative terminal as an incorrect polarisation will break down the insulating oxide layer and permanent damage may result.

All polarised electrolytic capacitors have their polarity clearly marked with a negative sign to indicate the negative terminal and this polarity must be followed.

Electrolytic Capacitors are generally used in DC power supply circuits due to their large capacitance's and small size to help reduce the ripple voltage or for coupling and decoupling applications. One main disadvantage of electrolytic capacitors is their relatively low voltage rating and due to the polarisation of electrolytic capacitors, it follows then that they must not be used on AC supplies. Electrolytic's generally come in two basic forms; Aluminium Electrolytic Capacitors and Tantalum Electrolytic Capacitors.

Aluminium & Tantalum Electrolytic Capacitor

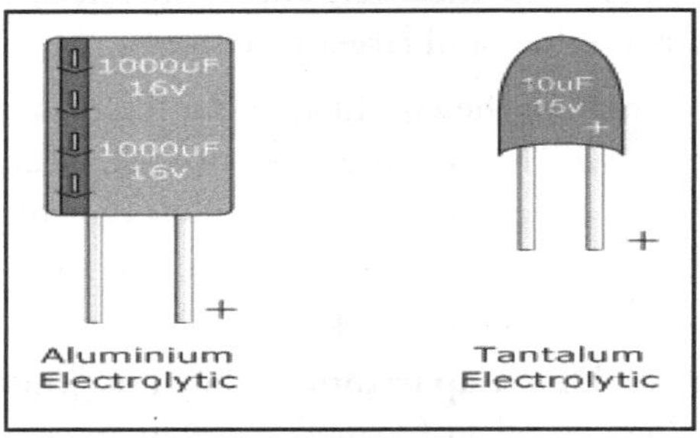

Application of Capacitors

1. ***Filter Applications:*** Combined with resistors, capacitors are often used as the main element of frequency selective filters. The available filter designs and topologies are numerous and can be tailored for frequency and performance by selecting the proper component values and quality. Some of the types of filter designs include:
   - High Pass Filter (HPF)
   - Low Pass Filter (LPF)
   - Band Pass Filter (BPF)

- Band Stop Filter (BSF)
- Notch Filter
- All Pass Filter
- Equalization Filter

2. ***Decoupling/By-Pass Capacitors:*** Capacitors play a critical role in the stable operation of digital electronics by protecting sensitive microchips from noise on the power signal which can cause anomalous behaviours. Capacitors used in this application are called decoupling capacitors and should be placed as close as possible to each microchip to be most effective, as all circuit traces act as antennas and will pick up noise from the surrounding environment. Decoupling and by-pass capacitors are also used in any area of a circuit to reduce the overall impact of electrical noise.
3. ***Coupling or DC Blocking Capacitors:*** Since capacitors have the ability to pass AC signals while blocking DC, they can be used to separate the AC and DC components of a signal. The value of the capacitor does not need to be precise or accurate for coupling, but it should be a high value as the reactance of the capacitor drives the performance in coupling applications.
4. ***Snubber or Transient suppression Capacitors:*** In circuits where a high inductance load is driven, such as a motor or transformer, large transient power spikes can occur as the energy stored in the inductive load is suddenly discharged which can damage components and contacts. Applying a capacitor can limit, or snub, the voltage spike across the circuit, making operation safer and the circuit more reliable. In lower power circuits, using a snubbing technique can be used to prevent spikes from creating undesirable radio frequency interference (RFI) which can cause anomalous behaviour in circuits and cause difficulty in gaining product certification and approval.
5. ***Pulsed Power Capacitors:*** At their most basic, capacitors are effectively tiny batteries and offer unique energy storage capabilities beyond those of chemical reaction batteries. When lots of power is required in a short period of time, large capacitors and banks of capacitors are a superior option for many applications. Capacitor banks are used to store energy for applications such as pulsed lasers, radars, particle accelerators, and railguns. A common application of the pulsed power capacitor is in the flash on a disposable camera which is charged up then rapidly discharged through the flash, providing a large pulse of current.
6. ***Resonant or Tuned Circuit Applications:*** While resistors, capacitors and inductors can be used to make filters, certain combinations can also result in resonance amplifying the input signal. These circuits are used to amplify signals at the resonant frequency, create high voltage from low voltage inputs, as oscillators, and as tuned filters. In resonant circuits, care must be taken to select components that can survive the voltages that the components see across them or they will quickly fail

7. ***Capacitive Sensing Application:*** Capacitive sensing has recently become a common feature in advanced consumer electronics devices, although capacitive sensors have been used for decades in a variety of applications for position, humidity, fluid level, manufacturing quality control and acceleration. Capacitive sensing works by detecting a change in the capacitance of the local environment through a change in the dielectric, a change in the distance between the plates of the capacitor, or a change in the area of a capacitor.

CHAPTER

6

Transformers

WORKING PRINCIPLE OF A TRANSFORMER, SPECIFICATIONS OF A TRANSFORMER, STEP-UP, STEP DOWN AND ISOLATION TRANSFORMERS

The Voltage Transformer can be thought of as an electrical component rather than an electronic component. A transformer basically is very simple static (or stationary) electro-magnetic passive electrical device that works on the principle of Faraday's law of induction by converting electrical energy from one value to another.

The transformer does this by linking together two or more electrical circuits using a common oscillating magnetic circuit which is produced by the transformer itself. A transformer operates on the principals of "electromagnetic induction", in the form of Mutual Induction.

Mutual induction is the process by which a coil of wire magnetically induces a voltage into another coil located in close proximity to it. Then we can say that transformers work in the "magnetic domain", and transformers get their name from the fact that they "transform" one voltage or current level into another.

Transformers are capable of either increasing or decreasing the voltage and current levels of their supply, without modifying its frequency, or the amount of electrical power being transferred from one winding to another via the magnetic circuit.

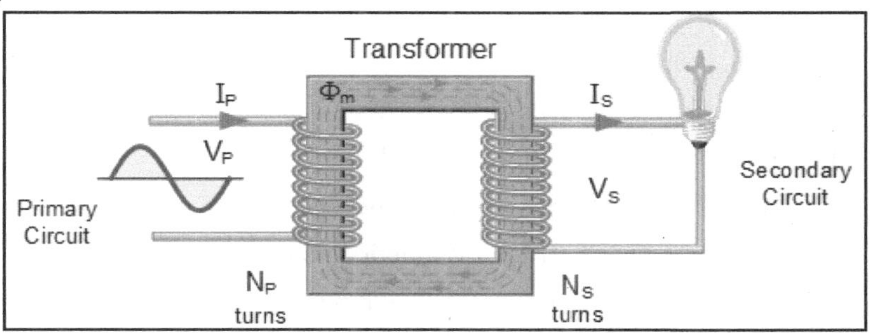

Figure 6.1: Transformer

In other words, for a transformer there is no direct electrical connection between the two coil windings, thereby giving it the name also of an **Isolation Transformer**. Generally, the primary winding of a transformer is connected to the input voltage supply and converts or transforms the electrical power into a magnetic field. While the job of the secondary winding is to convert this alternating magnetic field into electrical power producing the required output voltage as shown.

A single phase voltage transformer basically consists of two electrical coils of wire, one called the "Primary Winding" and another called the "Secondary Winding". We will define the "primary" side of the transformer as the side that usually takes power, and the "secondary" as the side that usually delivers power. In a single-phase voltage transformer the primary is usually the side with the higher voltage.

These two coils are not in electrical contact with each other but are instead wrapped together around a common closed magnetic iron circuit called the "core". This soft iron core is not solid but made up of individual laminations connected together to help reduce the core's losses.

The two coil windings are electrically isolated from each other but are magnetically linked through the common core allowing electrical power to be transferred from one coil to the other. When an electric current passed through the primary winding, a magnetic field is developed which induces a voltage into the secondary winding as shown.

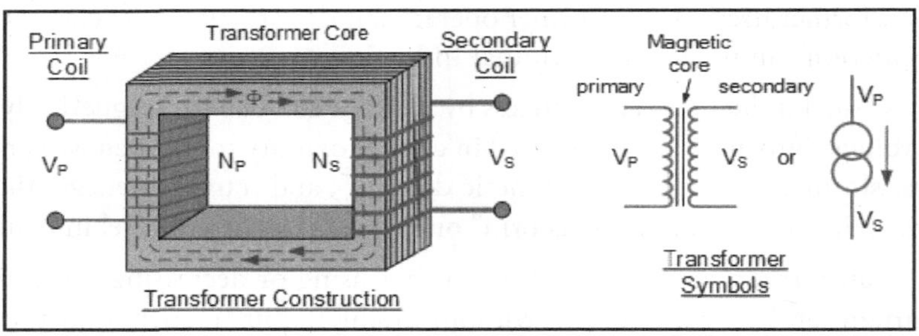

Figure 6.2: Transformer Construction

- Where:
- VP - is the Primary Voltage
- VS - is the Secondary Voltage
- NP - is the Number of Primary Windings
- NS - is the Number of Secondary Windings
- Φ (phi) - is the Flux Linkage (Magnetic linkage is also known as Flux linkage)

Notice that the two coil windings are not electrically connected but are only linked magnetically. A single-phase transformer can operate to either increase or decrease the voltage applied to the primary winding.

Step-up transformer: When a transformer is used to "increase" the voltage on its secondary winding with respect to the primary, it is called a Step-up transformer.

Step-down transformer: When a transformer is used to "decrease" the voltage on the secondary winding with respect to the primary it is called a Step-down transformer.

Isolation Transformer: However, a third condition exists in which a transformer produces the same voltage on its secondary as is applied to its primary winding. In other words, its output is identical with respect to voltage, current and power transferred. This type of transformer is called an "Impedance Transformer" and is mainly used for impedance matching or the isolation of adjoining electrical circuits.

How the voltage changes in Primary and Secondary side: The difference in voltage between the primary and the secondary windings is achieved by changing the number of coil turns in the primary winding (NP) compared to the number of coil turns on the secondary winding (NS).

As the transformer is basically a linear device, a ratio now exists between the number of turns of the primary coil divided by the number of turns of the secondary coil. This ratio, called the ratio of transformation, more commonly known as a transformers "turns ratio", (TR). This turns ratio value dictates the operation of the transformer and the corresponding voltage available on the secondary winding.

It is necessary to know the ratio of the number of turns of wire on the primary winding compared to the secondary winding. The turns ratio, which has no units, compares the two windings in order and is written with a colon, such as 3:1 (3-to-1). This means in this example, that if there are 3 volts on the primary winding there will be 1 volt on the secondary winding, 3 volts-to-1 volt. Then we can see that if the ratio between the number of turns changes the resulting voltages must also change by the same ratio, and this is true.

Transformers are all about "ratios". The ratio of the primary to the secondary, the ratio of the input to the output, and the turns ratio of any given transformer will be the same as its voltage ratio. In other words for a transformer: "turns ratio = voltage ratio". The actual number of turns of wire on any winding is generally not important, just the turns ratio and this relationship is given as:

A Transformers Turns Ratio T.R

T.R= N_P/N_S = Number of Turns of Primary Coil/ Number of Turns of Secondary Coil=V_P/V_S=Primary Voltage/Secondary Voltage

Note that the order of the numbers when expressing a transformers *turns ratio* value is very important as the turns ratio 3:1 expresses a very different transformer relationship and output voltage than one in which the turns ratio is given as: 1:3.

Transformer Basics Example No1- Step Down Transformer

A voltage transformer has 3000 turns of wire on its primary coil and 1000 turns of wire for its secondary coil. What will be the turns ratio (TR) of the transformer.

T.R= N_P/N_S = Number of Turns of Primary Coil/ Number of Turns of Secondary Coil = 3000/1000= 3:1=V_P/V_S=Primary Voltage/Secondary Voltage

This ratio of 3:1 (3-to-1) simply means that there are three primary windings for every one secondary winding. As the ratio moves from a larger number on the left to a smaller number on the right, the primary voltage is therefore stepped down in value as shown.

Thus if a Voltage of 240 V ac is applied on the primary or Input side there will be a Voltage of 240/3=80 V ac on the Secondary or Output side. Thus it is a Step Down Transformer.

$$TR= V_P/V_S$$

Then the main purpose of a transformer is to transform voltages at preset ratios and we can see that the primary winding has a set amount or number of windings (coils of wire) on it to suit the input voltage. If the secondary output voltage is to be the same value as the input voltage on the primary winding, then the same number of coil turns must be wound onto the secondary core as there are on the primary core giving an even turns ratio of 1:1 (1-to-1). In other words, one coil turn on the secondary to one coil turn on the primary.

Transformer Basics Example No2- Step Up Transformer

We now reverse the situation. A voltage transformer has 1000 turns of wire on its primary coil and 3000 turns of wire for its secondary coil. What will be the turns ratio (TR) of the transformer.

T.R= N_P/N_S = Number of Turns of Primary Coil/ Number of Turns of Secondary Coil = 1000/3000= 1:3

This ratio of 1:3 (1-to-3) simply means that there are three secondary windings for every one primary winding. As the ratio moves from a smaller number on the left to a larger number on the right, the primary voltage is therefore stepped up in value as shown.

Thus if a Voltage of 12 V ac is applied on the primary or Input side there will be a Voltage of 36 V ac on the Secondary or Output side. Thus it is a Step Up Transformer.

Specifications of a Transformer

The following are the main specifications of a Transformer:

1. Type- Step up/ Step Down.
2. Material- Coils- Copper
3. Material- Core - Laminated Iron.

4. Turn Ratio- ex- 3:1
5. Primary Voltage- 240 Volts.
6. Secondary Voltage- ex 80 Volts.
7. VA Rating= This is the product of Voltage and Amperes on any side. This will be constant on both the sides. Thus for a Primary Transformer of TR= 3:1 the voltage on the secondary side will decrease by three times where as the current will increase three times. Thus 240 VA (This means on the primary side you have 1 A and on the secondary side you have 3 A). If we apply 240 V and the maximum load is 1 A then in the secondary side we will have a voltage of 80 V and the current will be 3 A.
8. Frequency- 50 Hz.
9. Maximum Temperature Rise- Ex. 115 deg C
10. Maximum Insulation Tolerance- Ex. Upto 180 deg

These are the basic specifications.

Brainstorming- What is the output in the secondary for TR=3:1 if we apply 240 V dc??

Answer: There will be no output on the secondary side. This is because the transformer works on the principle of Electromagnetic Mutual Induction , which requires a voltage source that changes with time(alternating source) . Thus since a DC Supply is unidirectional , the transformer can't work.

Also please note that as transformers require an alternating magnetic flux to operate correctly, transformers cannot therefore be used to transform or supply DC voltages or currents, since the magnetic field must be changing to induce a voltage in the secondary winding. In other words, transformers DO NOT operate on steady state DC voltages, only alternating or pulsating voltages.

If a transformers primary winding was connected to a DC supply, the inductive reactance of the winding would be zero as DC has no frequency, so the effective impedance of the winding will therefore be very low and equal only to the resistance of the copper used. Thus the winding will draw a very high current from the DC supply causing it to overheat and eventually burn out, because as we know I = V/R.

Electrical Power in a Transformer

Another one of the transformer basics parameters is its power rating. The power rating of a transformer is obtained by simply multiplying the current by the voltage to obtain a rating in Volt-amperes, (VA). Small single phase transformers may be rated in volt-amperes only, but much larger power transformers are rated in units of Kilo volt-amperes, (kVA) where 1 kilo volt-ampere is equal to 1,000 volt-amperes, and units of Mega volt-amperes, (MVA) where 1 mega volt-ampere is equal to 1 million volt-amperes.

In an ideal transformer (ignoring any losses), the power available in the secondary winding will be the same as the power in the primary winding, they are constant wattage devices and do not change the power only the voltage to current ratio. Thus, in an ideal transformer the Power Ratio is equal to one (unity) as the voltage, V multiplied by the current, I will remain constant.

That is the electric power at one voltage/current level on the primary is "transformed" into electric power, at the same frequency, to the same voltage/current level on the secondary side. Although the transformer can step-up (or step-down) voltage, it cannot step-up power. Thus, when a transformer steps-up a voltage, it steps-down the current and vice-versa, so that the output power is always at the same value as the input power. Then we can say that primary power equals secondary power, (PP = PS).

Power in a Transformer

$$\text{Power}_{\text{Primary}} = \text{Power}_{\text{Secondary}}$$

$$P_{(PRIM)} = P_{(SEC)} = V_P I_P \cos\theta_P = V_S I_S \cos\theta_S$$

Where: Φ_P is the primary phase angle and Φ_S is the secondary phase angle.

Note that since power loss is proportional to the square of the current being transmitted, that is: I^2R, increasing the voltage, let's say doubling (×2) the voltage would decrease the current by the same amount, (÷2) while delivering the same amount of power to the load and therefore reducing losses by factor of 4. If the voltage was increased by a factor of 10, the current would decrease by the same factor reducing overall losses by factor of 100.

Transformer Basics – Efficiency

A transformer does not require any moving parts to transfer energy. This means that there are no friction or windage losses associated with other electrical machines. However, transformers do suffer from other types of losses called "copper losses" and "iron losses" but generally these are quite small.

Copper losses, also known as I2R loss is the electrical power which is lost in heat as a result of circulating the currents around the transformers copper windings, hence the name. Copper losses represents the greatest loss in the operation of a transformer. The actual watts of power lost can be determined (in each winding) by squaring the amperes and multiplying by the resistance in ohms of the winding (I2R).

Iron losses, also known as hysteresis is the lagging of the magnetic molecules within the core, in response to the alternating magnetic flux. This lagging (or out-of-phase) condition is due to the fact that it requires power to reverse magnetic molecules; they do not reverse until the flux has attained sufficient force to reverse them.

Their reversal results in friction, and friction produces heat in the core which is a form of power loss. Hysteresis within the transformer can be reduced by making the core from special steel alloys.

The intensity of power loss in a transformer determines its efficiency. The efficiency of a transformer is reflected in power (wattage) loss between the primary (input) and secondary (output) windings. Then the resulting efficiency of a transformer is equal to the ratio of the power output of the secondary winding, P_S to the power input of the primary winding, P_P and is therefore high.

An ideal transformer is 100% efficient because it delivers all the energy it receives. Real transformers on the other hand are not 100% efficient and at full load, the efficiency of a transformer is between 94% to 96% which is quiet good. For a transformer operating with a constant voltage and frequency with a very high capacity, the efficiency may be as high as 98%. The efficiency, η of a transformer is given as:

Transformer Efficiency

$$\text{efficiency, } \eta = \frac{\text{Output Power}}{\text{Input Power}} \times 100\%$$

$$= \frac{\text{Input Power - Losses}}{\text{Input Power}} \times 100\%$$

$$= \frac{\text{Losses}}{\text{Input Power}} \times 100\%$$

where: Input, Output and Losses are all expressed in units of power.

Generally when dealing with transformers, the primary watts are called "volt-amps", VA to differentiate them from the secondary watts. Then the efficiency equation above can be modified to:

$$\text{Efficiency, } \eta = \frac{\text{Secondary Watts (Output)}}{\text{Primary VA (Input)}}$$

It is sometimes easier to remember the relationship between the transformers input, output and efficiency by using pictures. Here the three quantities of VA, W and η have been superimposed into a triangle giving power in watts at the top with volt-amps and efficiency at the bottom. This arrangement represents the actual position of each quantity in the efficiency formulas.

Transformer Efficiency Triangle

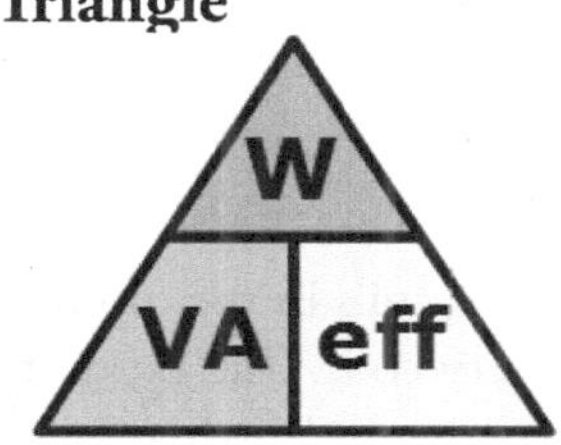

and transposing the above triangle quantities gives us the following combinations of the same equation:

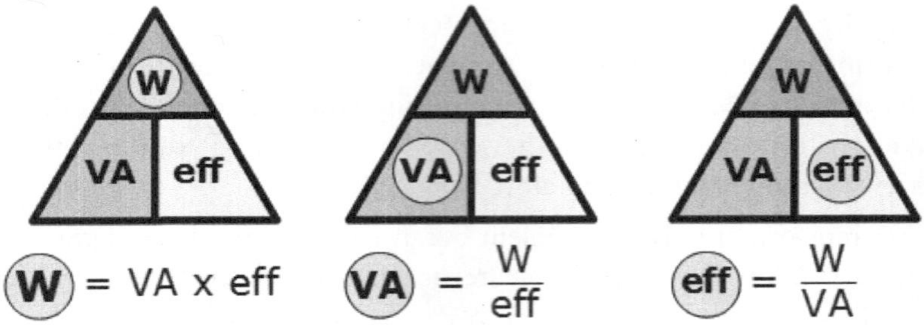

Then, to find Watts (output) = VA x eff., or to find VA (input) = W/eff., or to find Efficiency, eff. =W/VA, etc.

Assessment Questions:

1. What is a Transformer?

Ans: A Transformer changes the voltage level (or current level) on its input winding to another value on its output winding using a magnetic field.

2. What is the principle of operation of a Transformer?

Ans.: A transformer consists of two electrically isolated coils and operates on Faraday's principal of "mutual induction", in which an EMF is induced in the transformers secondary coil by the magnetic flux generated by the voltages and currents flowing in the primary coil winding.

3. How are the Primary and Secondary windings generally connected?

Ans.: Both the primary and secondary coil windings are wrapped around a common soft iron core made of individual laminations to reduce eddy current and power losses. The primary winding of the transformer is connected to the AC power source which must be sinusoidal in nature, while the secondary winding supplies power to the load.

4. Give a block diagram of a Transformer?

Ans.: We can represent the transformer in block diagram form as follows

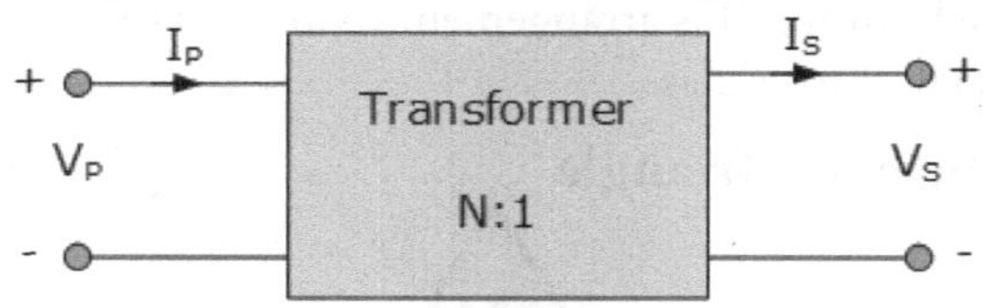

5. What is Turnover Ratio?

Ans.: This is the ratio of number of coils in the input i.e Primary Coil to the number of coils in the output i.e Secondary coils. The ratio of the transformers primary and secondary windings with respect to each other produces either a step-up voltage transformer or a step-down voltage transformer with the ratio between the number of primary turns to the number of secondary turns being called the "turns ratio" or "transformer ratio".

6. What is Step Up Transformer?

Ans.: If the Turnover Ratio is less than unity, n < 1 then NS is greater than NP and the transformer is classed as a step-up transformer. This increase the output voltage (and decreases the output current).

7. What is a Step down transformer.

Ans.: What is Step down Transformer- If the Turnover Ratio is greater than unity, n > 1, that is NP is greater than NS, the transformer is classed as a step-down transformer This decreases the output voltage (and increases the output current)

8. Can Step Down transformer be used as Step Up Transformer.? If Yes How?

Ans.: Yes, a single phase step-down transformer can also be used as a step-up transformer simply by reversing its connections and making the low voltage winding its primary, and vice versa as long as the transformer is operated within its original VA design rating.

9. What is an Isolation Transformer?

Ans.: If the turns ratio is equal to unity, n = 1 then both the primary and secondary have the same number of windings, therefore the voltages and currents are the same for both windings. This type of transformer is classed as an isolation transformer as both the primary and secondary windings of the transformer have the same number of volts per turn.

10. What is efficiency of a Transformer.?

Ans.: The efficiency of a transformer is the ratio of the power it delivers to the load to the power it absorbs from the supply. In an ideal transformer there are no losses so no loss of power then Pin = Pout.

CHAPTER

7

Electronics Active Components and Testing

BASIC ELECTRONICS ACTIVE COMPONENTS, TESTING OF COMPONENTS

Experiment no: 1

Name of Experiment: Testing of active &passive component with the help of C.R.O and digital multimeter

Aim: To study the function of digital multimeter & C.R.O & testing various components like passive and active components such as resistance, capacitance , inductance, classical diodes, zener diodes, photodiode, Light Emitting Diodes, BJT, JFET and MOSFET.

Theory: In component testing process, we test the continuity and the I-V characteristics of various component like resistor, capacitor, diode and transistors.

Wires and Connectors:we use the following wires:

1. Single Strand wire of SWG 22 or SWG23. SWG22 is preferred because it snugly ts the solder-less, spring loaded contact points in the breadboard. SNUGLY means neither loose nor tight. SWG is Standard Wire Guage. 22 means diameter ϕ = (1/22) inch. These are used as hookup wire for making connections on the breadboard.
2. Multi strand wires are exible wires used as connecting wires between the Power supply and the bus of the circuit. Bus are lines which act as GROUND/EARTH, lines which act as (+)ve bus or as (-)ve bus. The guage of multi-strand wire can be SWG 7/36 or SWG 14/36. SWG 14/36 means 14 strands and 1/36 inch in dia. These multi strand wires have banana plug on one end and crocodile clip on the other. By banana plug they are tied on the banana socket on the power supply and by crocodile clip it is clipped on the hook up wire connected in the bread board.

3. There are various types of wires. There are two cores or three cores shielded wires. In one sheath there are two or three insulated SWG14/36 wires. The sheath can be shielded in wire-mesh shields. This prevents Electro-magnetic interference and pick-up. There are unshielded wires with two and three cores.
4. There are coaxial cables which are used as connectors . A coaxial cable with BNC Connector on one end and crocodile clips on the other are used for feeding the signal from Function Generators to the circuit under test. A coaxial cable with BNC Connector on one end and oscilloscope probes or crocodile clips on the other are used for feeding the signal from the circuit under test to the oscilloscope.

Procedure:

- Set up the component in the breadboard which has to be tested & probed by the mulimiter as well as C.R.O.
- First of all we calculate the theoretical value of resistor through colour coding and than obtain practical value through multimeter.
- Next we nd the I-V characteristic curve i.e forward and reverse characteristics of diode with the help of C.R.O.

Resistances:

1. ***Carbon Film Reistances:*** There is a carbon lm deposited on the former. These are the cheapest and most widely used resistances. The resistances are given in colour codes.

   Colour code is Gray, White, Violet ,Blue ,Green, Yellow, Orange, Red Brown and Black. These stand for 9,8,7,6,5,4,3,2,1 and 0 respectively.

   The rest two bands on the left give tens and units. The third band gives the exponent to the base 10.

   Fourth band gives the tolerance. Golden Band gives 5% and Silver Band gives 10% tolerance.

S.No	1st,2nd,3rd and 4th band of colours	Nominal Value (Ω)	Practical Value (Ω)	Tolerance (± %)	Percentage Error
1	Yellow, Violet, Orange, Silver	47k	47.2k	10	0.423
2	Red, Violet, Red, Gold	2.7k	2.748k	5	1.746

   The resistances have power rating and tolerance. Power rating tell the maximum power dissipation permitted. These are generally 1/4W or 1/2W rating in carbon lm resistances.

   The fourth band tells the tolerance. Zero Tolerance resistance is a precision resistance. Less than 1% tolerance is close tolerance component. More than 5% tolerance is a wide tolerance component.

2. Thin Film Resistances: Thin Film Resistances are made of Nichrome which is 80% Nickel and 20% Chromium alloy. These are of 0.1% tolerance and the temperature coecient of resistance(t.c.r) is 100 parts per million per degree centigrade (PPM/°C). These are also called precision resistances.
3. CERMETS are lm resistances where ceramic and metal lms are used. These are also of 0.1% tolerance and zero t.c.r. Ceramic has a negative t.c.r. and Metal has positive t.c.r. Hence a proper combination of Ceramic and Metal gives zero t.c.r. These are truly precision components used in Instrumentation.
4. ***Wire Wound Resistances:*** These are power resistances which can go up to 100W maximum power dissipation. These are green in colour and robust in appearance.
5. ***Variable Resistances:*** These are carbon track potentiometers and Rheostats. Carbon track resistances are used in Electronics Lab and are of less than 1W rating. Rheostats are used in Electrical Engineering Labs and can go up to 1000W. Rheostat are made of Cu/Ni or Ni/Cr alloy of low t.c.r..

In Carbon Track Potentiometer there is a carbon track on which a wiper sweeps from one end to the other. The two terminals connected to the two ends of the carbon track give a xed resistance while the resistance between one of the end terminals and the terminal connected to the wiper is a variable resistance.

Capacitor:

1. ***Ceramic Disc Capacitances:*** These are the cheapest and most widely used capacitances. The values are given as 103. 103 means 10×103 pF (pF is called PUF). This is 10-8 F=0.01μF.
2. Paper/Mica Capacitances.
3. ***Polyesterene Capacitances:*** These are low loss angle high quality capacitances.
4. ***Electrolytic Capacitances:*** The large magnitude of capacitances more than 1μF up to 1000μF can be realized by electrolytic Capacitances. These are polar capacitances. The polarity of the DC Voltage applied should be correct so that the electrolyte between the electrodes act like dielectric. If the polarity is reversed then it becomes a conductor.

4. ***Variable Capacitances:*** Gang Capacitances- these are air-gap capacitances used as tuning capacitances in Radio-Broadcast Receivers. Trimmers and Padders are used for alignment of the tuned circuits in Radio Receivers.

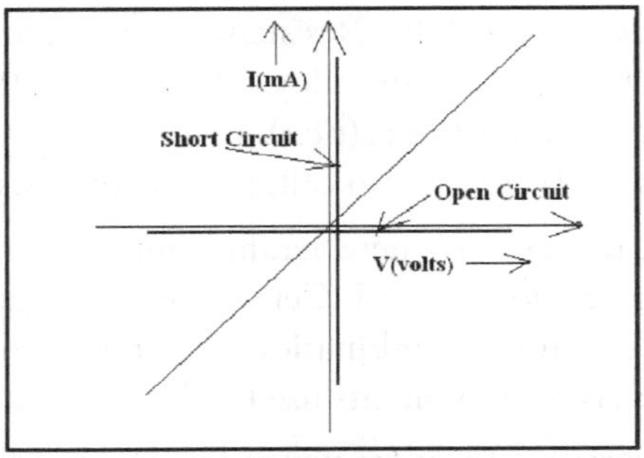

Figure 7.1: VI Characterstics of a resistor

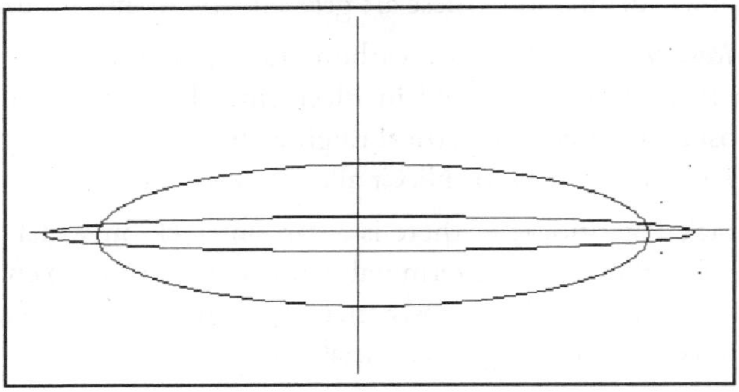

Figure 7.2: IV Characteristics of a Capacitor

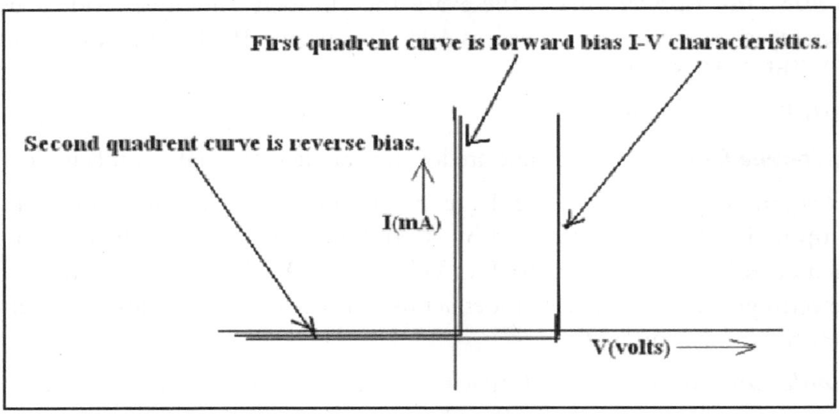

Figure 7.3: IV Characteristics of a Diode. In the First Quadrant we see the forward bias and cut off voltage.

The cut off Voltage For Germenium diode it is 0 Volts whereas for Silicon Diode it is 0.5 V. But for LED it is 1.2V as it is a compound semiconductor.

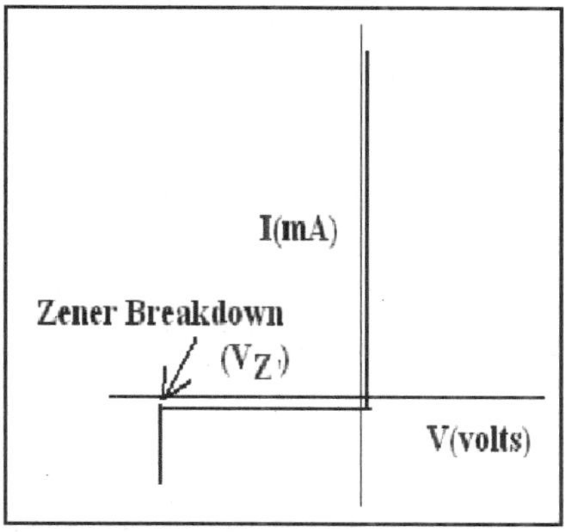

Figure 7.4: IV Characteristics of Zener Diode.

Zener Diode operates in Reverse Bias and the voltage regulation occurs because of zener breakdown.

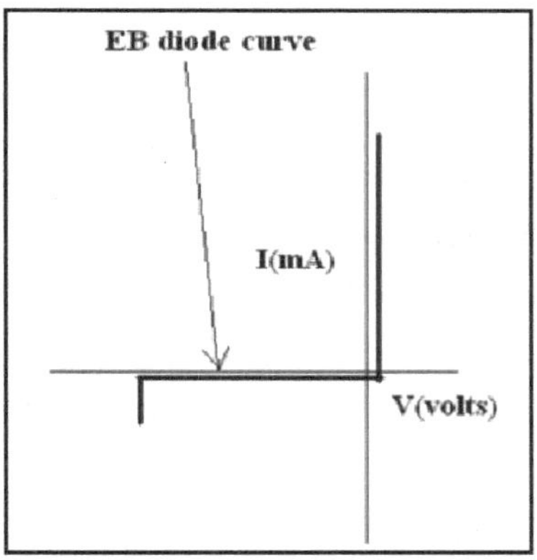

Figure 7.5: EB Diode IV Characteristics EB Diode behaves like a Zener Diode. Because of heavy doping

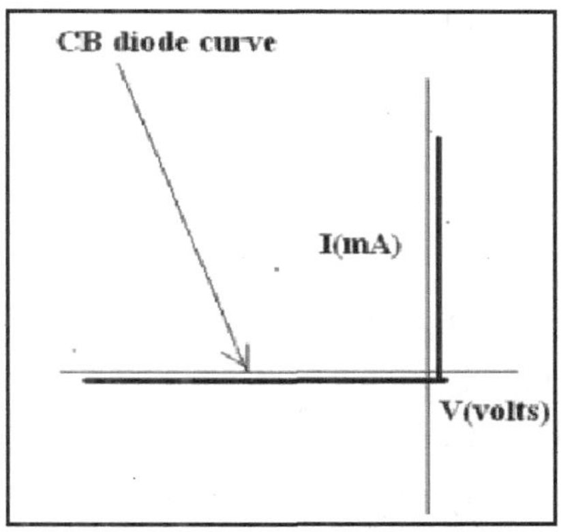

Figure 7.6: CB Diode IV Characteristics CB diode behaves like classical diode. Because of low doping

Basic Power Supply:

Parts of a Power Supply- A DC Power Supply Unit (commonly called a PSU) deriving power from the AC mains (line) supply performs a number of tasks:

1. It changes (in most cases reduces) the level of supply to a value suitable for driving the load circuit.
2. It produces a DC supply from a pure AC wave.
3. It prevents any AC from appearing at the supply output.
4. It will ensure that the output voltage is kept at a constant level, independent of changes in:

a. The AC supply voltage at the supply input.

b. The Load current drawn from the supply output.

c. Temperature.

The following is the basic block diagram of a Power Supply:

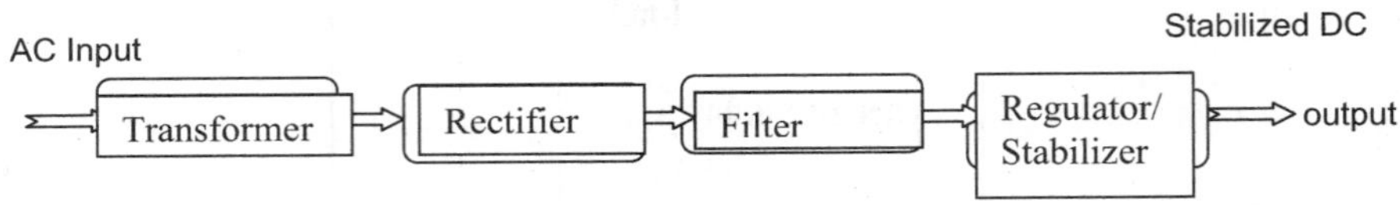

Transformer stage:

The Transformer

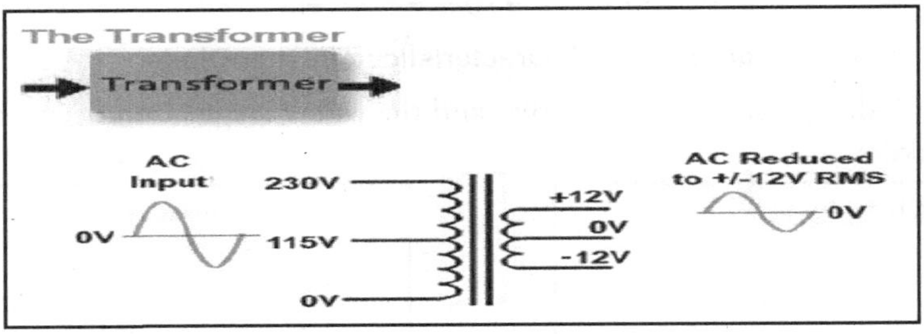

Figure 7.7: Typical Input Transformer

In a basic power supply the input power transformer has its primary winding connected to the mains

(line) supply. A secondary winding, electro-magnetically coupled but electrically isolated from the

primary is used to obtain an AC voltage of suitable amplitude, and after further processing by the

PSU, to drive the electronics circuit it is to supply.

The transformer stage must be able to supply the current needed. If too small a transformer is used,

it is likely that the power supply's ability to maintain full output voltage at full output current will be impaired. With too small a transformer, the losses will increase dramatically as full load is placed on the transformer.

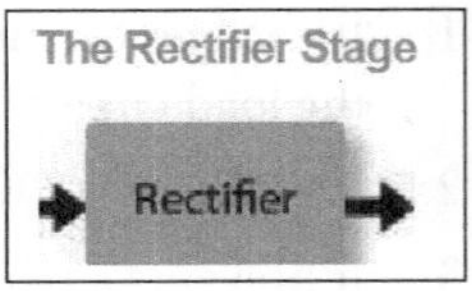

As the transformer is likely to be the most costly item in the power supply unit, careful consideration must be given to balancing cost with likely current requirement. There may also be a need for safety devices such as thermal fuses to disconnect the transformer if overheating occurs, and electrical isolation between primary and secondary windings, for electrical safety.

Three types of silicon diode rectifier circuit may be used, each having a different action in the way that the AC input is converted to DC. These differences are illustrated in Figures below.

WORKING OF HALF WAVE RECTIFIER

Half Wave Rectification

A single silicon diode may be used to obtain a DC voltage from the AC input as shown in Figure. This system is cheap but is only suitable for fairly non-demanding uses. The DC voltage produced by the single diode is less than with the other systems, limiting the efficiency of the power supply, and the amount of AC ripple left on the DC supply is generally greater.

The half wave rectifier conducts on only half of each cycle of the AC input wave, effectively blocking the other half cycle, leaving the output wave shown in Fig. below. As the average DC value of one half cycle of a sine wave is 0.637 of the peak value, the average DC value of the whole cycle after half wave rectification will be 0.637 divided by 2, because the average value of every alternate half cycle where the diode does not conduct, will of course be zero. This gives an output of:

$$Vpk \times 0.318$$

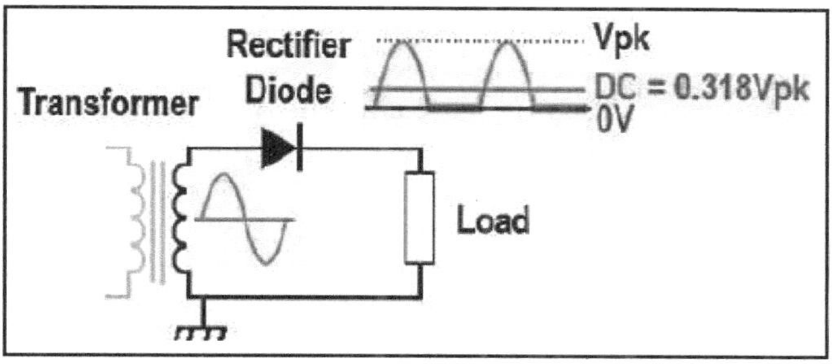

Figure .78: Half wave Rectifier

This figure is approximate, as the amplitude of the half cycles for which the diode conducts will also be reduced by about 0.6V due to the forward voltage drop (the depletion layer p.d.) of the silicon rectifier diode. This additional voltage drop may be insignificant when large voltages are rectified, but in low voltage power supplies where the AC from the secondary winding of the mains transformer may be only a few volts, this 0.6V drop across the diode junction may have to be compensated for, by having a slightly higher transformer secondary voltage.

Half wave rectification is not very efficient at producing DC from a 50Hz or 60Hz AC input. In addition the gaps between the 50 or 60Hz diode output pulses make it more difficult to remove the AC ripple remaining after rectification.

Full Wave Rectification

If a transformer with a centre tapped secondary winding is used, more efficient full wave rectification can be used. The centre-tapped secondary produces two anti-phase outputs, as shown in Fig below:

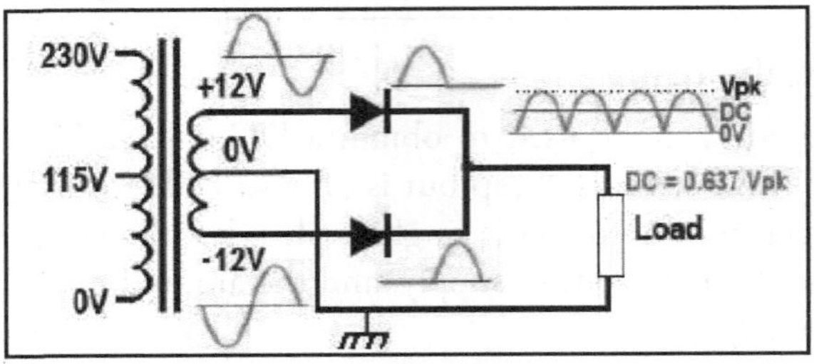

Figure 7.9: Full wave rectifier

If each of these outputs is 'half wave rectified' by one of the two diodes, with each diode conducting on alternate half cycles, two pulses of current occur at every cycle, instead of once per cycle in half wave rectification. The output frequency of the full wave rectifier is therefore twice the input frequency. This effectively provides twice the output voltage of the half wave circuit, Vpk x 0.637 instead of Vpk x 0.318 as the 'missing' half cycle is now rectified, reducing the power wasted in the half wave circuit. The higher output frequency also makes the smoothing of any remaining AC ripple easier.

Although this full wave design is more efficient than the half wave, it requires a centre tapped (and therefore more expensive) transformer.

The Bridge Rectifier

The full wave bridge rectifier uses four diodes arranged in a bridge circuit as shown in Fig. below to give full wave rectification without the need for a centre-tapped transformer. An additional advantage is that, as two diodes (effectively in series) are conducting at any one time, the diodes need only half the reverse breakdown voltage capability of diodes used for half and conventional full wave rectification. The bridge rectifier can be built from separate diodes or a combined bridge rectifier can be used.

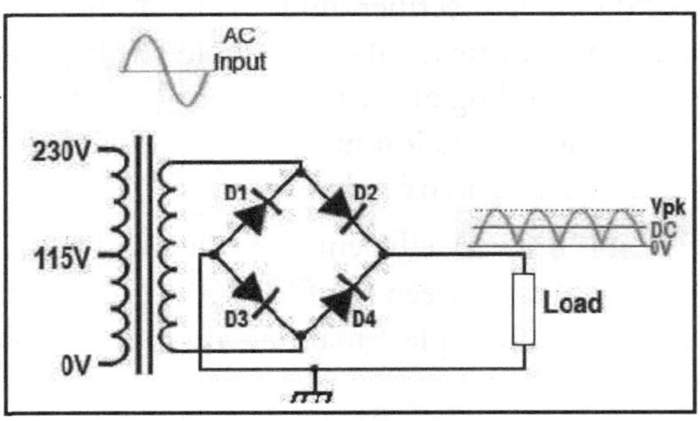

Figure 7.10: Bridge Rectifier

The current paths on positive and negative half cycles of the input wave are shown in Figures below . It can be seen that on each half cycle, opposite pairs of diodes conduct, but the current through the load remains in the same polarity for both half cycles.

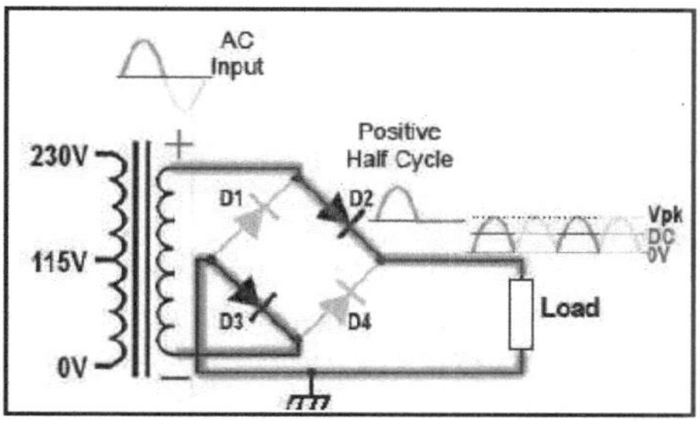

Figure 7.11: Current Flow on positive half cycle

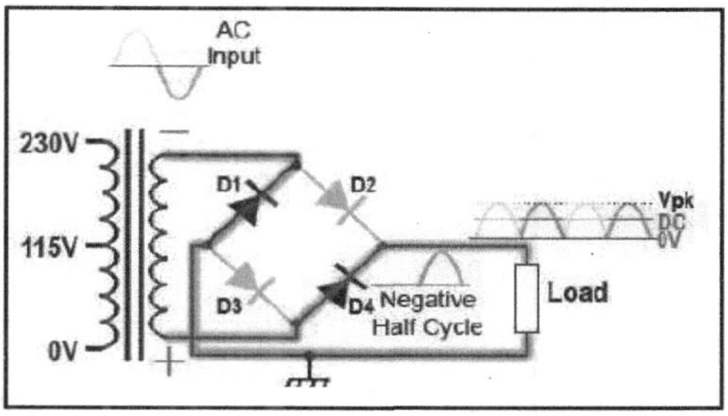

Figure 7.12: CurrentFlow on negative half cycle

Filter Components

A typical power supply filter circuit can be best understood by dividing the circuit into two parts, the reservoir capacitor and the low pass filter. Each of these parts contributes to removing the remaining AC pulses, but in different ways.

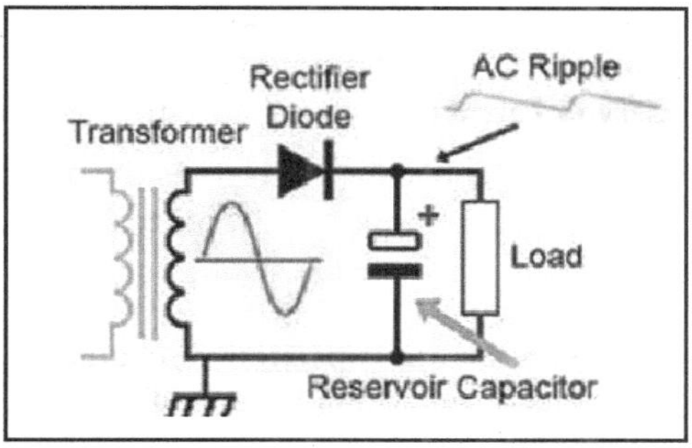

Figure 7.13: Reservoir Capacitor

Fig. above shows an electrolytic capacitor used as a reservoir capacitor, so called because it acts as a temporary storage for the power supply output current. The rectifier diode supplies current to charge a reservoir capacitor on each cycle of the input wave. The reservoir capacitor is a large electrolytic, usually of several hundred or even a thousand or more microfarads, especially in mains frequency PSUs. This very large value of capacitance is required because the reservoir capacitor, when charged, must provide enough DC to maintain a steady PSU output in the absence of an input current; i.e. during the gaps between the positive half cycles when the rectifier is not conducting.

The action of the reservoir capacitor on a half wave rectified sine wave is shown in Figure below. During each cycle, the rectifier anode AC voltage increases towards Vpk. At some point close to Vpk the anode voltage exceeds the cathode voltage, the rectifier conducts and a pulse of current flows, charging the reservoir capacitor to the value of Vpk.

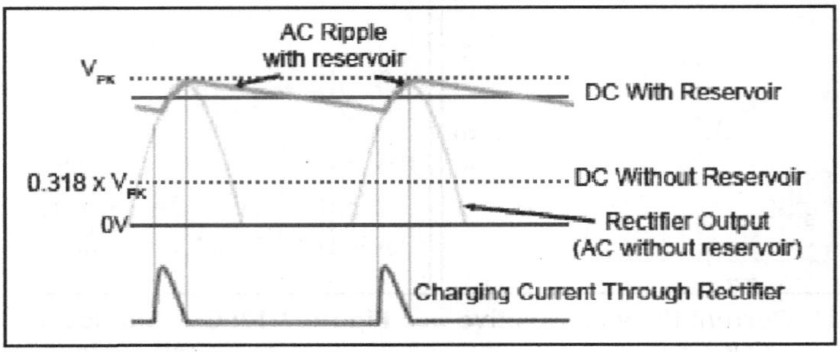

Figure 7.14: Reservoir capacitor action

Of course, even though the reservoir capacitor has large value, it discharges as it supplies the load,

and its voltage falls, but not by very much. At some point during the next cycle of the mains input,

Once the input wave passes Vpk the rectifier anode falls below the capacitor voltage, the rectifier becomes reverse biased and conduction stops. The load circuit is now supplied by the reservoir capacitor alone (hence the need for a large capacitor).

The rectifier input voltage rises above the vo l tage on the partly discharged capacitor and the reservoir is re-charged to the peak value Vpk again.

AC Ripple

The amount by which the reservoir capacitor discharges on each half cycle is determined by the current drawn by the load. The higher the load current, the more the discharge, but provided that the current drawn is not excessive, the amount of the AC present in the output is much reduced. Typically the peak-to-peak amplitude of the remaining AC (called ripple as the AC waves are now much reduced) would be no more than 10% of the DC output voltage.

The DC output of the rectifier, without the reservoir capacitor, is either 0.637 Vpk for full wave rectifiers, or 0.317 Vpk for half wave. Adding the capacitor increases the DC level of the output wave to nearly the peak value of the input wave.

To obtain the least AC ripple and the highest DC level it would seem sensible to use the largest reservoir capacitor possible. There is a snag however. The capacitor supplies the load current for most of the time (when the diode is not conducting). This current partly discharges the capacitor, so all of the energy used by the load during most of the cycle must be made up in the very short remaining time during which the diode conducts in each cycle.

The formula relating charge, time and current states that:

$$Q = It$$

The charge (Q) on a capacitor depends on the amount of current (I) flowing for a time (t). Therefore the shorter the charging time, the larger current the diode must supply to charge it. If the capacitor is very large, its voltage will hardly fall at all between charging pulses; this will produce a very small amount of ripple, but require very short pulses of much higher current to charge the reservoir capacitor. Both the input transformer and the rectifier diodes must be capable of supplying this current. This means using a higher current rating for the diodes and the transformer than would be necessary with a smaller reservoir capacitor. There is an advantage therefore in reducing the value of the reservoir capacitor, thereby allowing an increase in the ripple present, but this can be effectively removed by using a low pass filter and regulator stages between the reservoir capacitor and the load.

This effect of increasing reservoir size on diode and transformer current should be born in mind during any servicing operations; replacing the reservoir capacitor with a larger value than in the original design "to reduce mains hum" may seem like a good idea, but could risk damaging the rectifier diode and/or the transformer.

With full wave rectification the performance of the reservoir capacitor in removing AC ripple is significantly better than with half wave, for the same size of reservoir capacitor, the ripple is about half the amplitude of that in half wave supplies, because in full wave circuits, discharge periods are shorter with the reservoir capacitor being recharged at twice the frequency of the half wave design.

Low Pass Filters

Although a useable power supply can be made using only a reservoir capacitor to remove AC ripple, it is usually necessary to also include a low pass filter and/or a regulator stage after the reservoir capacitor to remove any remaining AC ripple and improve the stabilisation of the DC output voltage under variable load conditions.

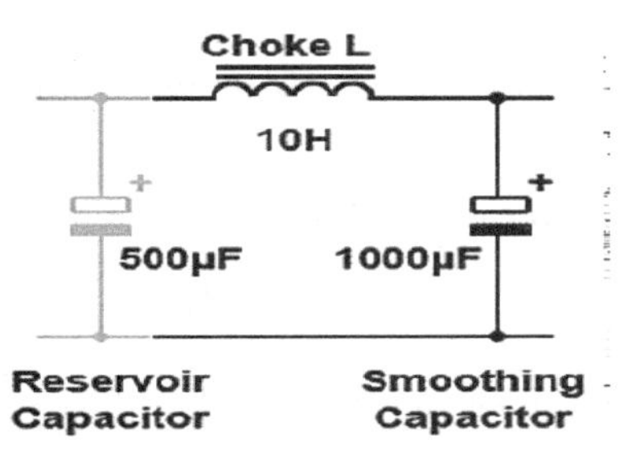

Figure 7.15: LC Filters

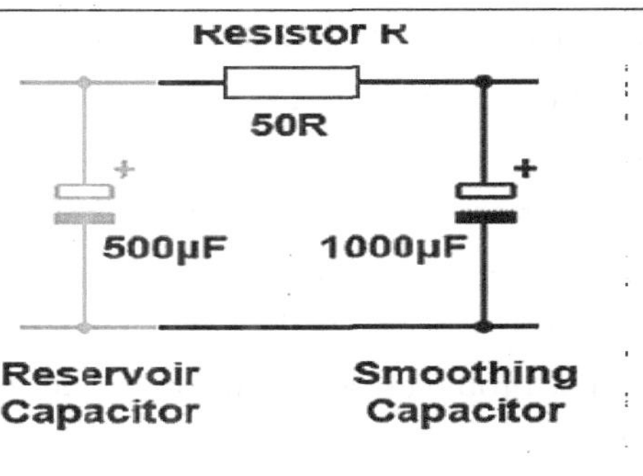

Figure 7.16: RC Filters

Either LC or RC low pass filters can be used to remove the ripple remaining after the reservoir capacitor. The LC filter shown in Fig. below is more efficient and gives better results than the RC filter shown in Figure but for basic power supplies, LC designs are less popular than RC, as the inductors needed for the filter to work efficiently at 50 to 120Hz need to be large and expensive laminated or toroidal core types. However modern designs using switch mode supplies, where any AC ripple is at much higher frequencies, much smaller ferrite core inductors can be used. The low pass filter passes low frequency, in this case DC (0Hz) and blocks higher frequencies, whether 50Hz or 120Hz in basic circuits or tens of kHz in switch mode designs. The reactance (XC) of the capacitor in the either of the filters is very low compared with the resistance of resistor R or the reactance of the choke XL at the ripple frequency. In RC designsthe resistance of R must be a fairly low value as the entire load current, maybe several amperes, must pass through it, generating a considerable amount of heat. A typical value would therefore be 50 ohms or less, and even at this value, a large wire wound resistor would normally need to be used. This limits the efficiency of the filter as the ratio between the resistance of R and the capacitor reactance will not be greater than about 25:1. This then would be the typical reduction ratio of the ripple amplitude. By including the low pass filter some voltage is lost across the resistor, but this disadvantage is offset by the better ripple performance than by using the reservoir capacitor alone.

The LC filter performs much better than the RC filter because it is possible to make the ratio between XC and XL much bigger than the ratio between XC and R. Typically the ratio in a LC filter could be 1:4000 giving much better ripple rejection than the RC filter. Also, since the DC resistance of the inductor in the LC filter is much less than the resistance of R in the RC filter, the problem of heat being generated by the large DC current is very much reduced in LC filters. With a combined reservoir capacitor and low pass filter it is possible to remove 95% or more of the AC ripple and obtain an output voltage of about the peak voltage of the input wave. A simple power supply consisting of only transformer, rectifier, reservoir and low pass filter however, does have some drawbacks.

The output voltage of the PSU tends to fall as more current is drawn from the output. This is due to:

a. The reservoir capacitor being discharged more on each cycle.

b. Greater voltage drop across the resistor or choke in the low pass filter as current increases.

These problems can be largely overcome by including a regulator stage at the power supply output.

Power supply Quiz

1. Refer to Fig 7.17. below. What is the function of block B?

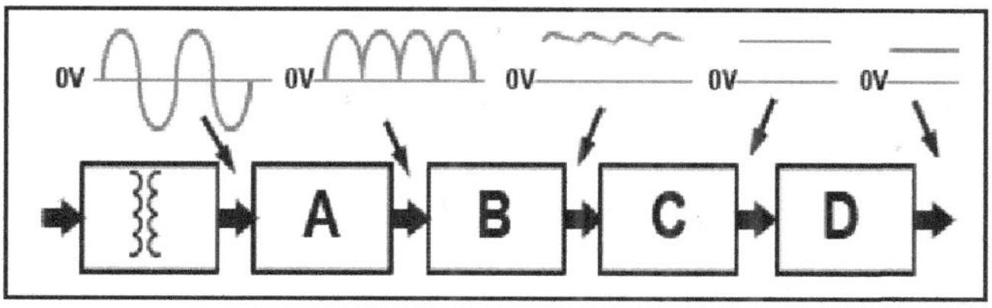

Figure 7.17

   a) Rectifier.
   b) Reservoir capacitor.
   c) Low pass filter.
   d) Regulator.

2. Refer to Fig-1 What is the function of block A?
   a) Transformer.
   b) Full wave rectifier.
   c) Bridge rectifier.
   d) Reservoir capacitor.

3. Refer to Figure. What will be the approximate value of the DC component of the waveform at the output of block A?
   a) VPK x 0.318
   b) VPK x 0.5
   c) VPK x 0.637
   d) VPK x 0.707

4. Refer to Figure 7.18 below. If input B is more positive than input A, which diodes will be conducting?

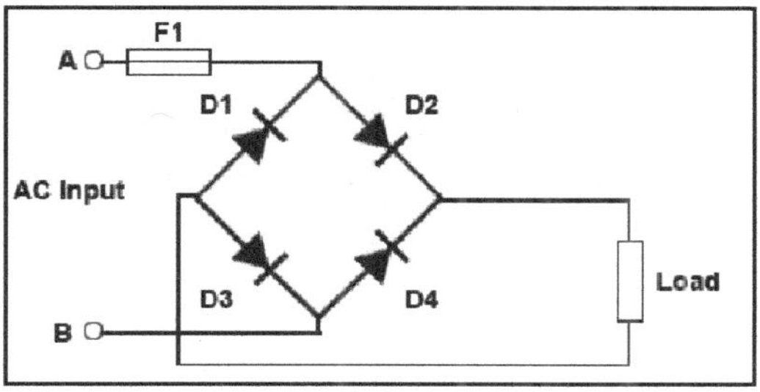

Figure 7.18

   a) D1 and D2.

b) D2 and D3

c) D1 and D4

d) D3 and D4

5. Refer to Fig 2. If D4 were to go short circuit, what would be the effect on the operation of the circuit?

a) A decrease in the current through D1.

b) Fuse F1 would blow.

c) A higher voltage across the load.

d) A larger peak current through D2 and D3.

6. What is the action of the reservoir capacitor in a basic power supply circuit?

a) To de-couple the DC component of the rectifier AC output.

b) To Increase the DC component and reduce the AC component of the AC wave.

c) To remove the DC component of the AC wave.

d) To regulate the AC wave.

7. Which of the following is an advantage of using a LC low pass filter rather than a RC low pass filter in a power supply?

a) The reactance of L will be much lower than the resistance of R at mains frequency

b) The reactance of L will be much higher than the resistance of R at mains frequency.

c) An inductor can dissipate more power than a resistor.

d) LC filters are less expensive than RC filters.

8. Refer to Figure 7.19 What is the power dissipated in R1?

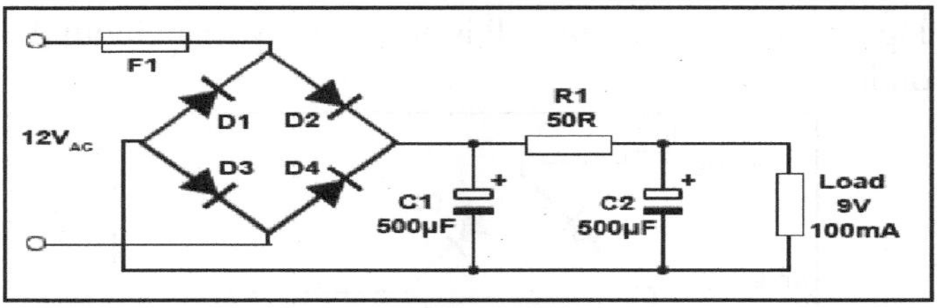

Figure 7.19

a) 5W

b) 2W

c) 500mW

d) 50mW

9. Refer to Fig -3. What will be the approximate value of DC across C1?
   a) 3.8V
   b) 7.6V
   c) 10.8V
   d) 14.5V
10. Refer to Fig -3. What is the reactance of C2 at the ripple frequency?
   a) 6.4
   b) 0.3
   c) 5.1
   d) 3.2

Answers

VOLTAGE REGULATOR CIRCUIT

Regulated Power Supply

Regulated power supply is an electronic circuit that is designed to provide a constant dc voltage of predetermined value across load terminals irrespective of ac mains fluctuations or load variations.

As shown in the figure, the two main parts of a regulated power supply are a simple power supply and a voltage regulating device. The power supply output is given as input to the voltage regulating device that provides the final output. The voltage output of the power supply remains constant irrespective of large variations in the input AC voltage or output load current.

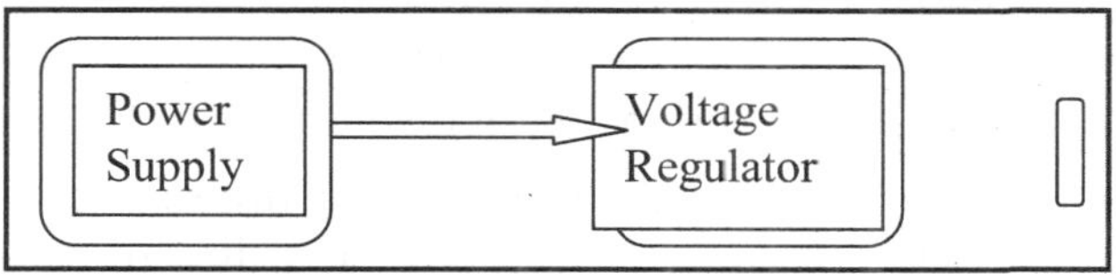

Given below is a circuit diagram of a regulated power supply circuit using a transistor series regulator as a regulating device. The input AC voltage (230 Voltas Vrms), is supplied to a transformer. The output will be a stepped down ac output appropriate for the desired dc output. This ac voltage is then given to a bridge rectifier to produce a full-wave rectified output. This is then given to a pi-filter circuit to produce a dc voltage. The filter output may have some ac voltage variations and ripples. This is further filtered using a regulating circuit whose output will be a constant dc voltage. This regulated dc voltage is then given to a voltage divider, which supplies the different dc voltages that may be needed for different electronic circuits.

The potential divider is a single tapped resistor connected across the output terminals of the supply. The tapped resistor may consist of two or three resistors connected in series across the supply. A bleeder resistor may also be employed as a potential divider.

Summary of Power Supply basics

Power supplies or power supply units, PSU, form an essential part of very many items of electronics equipment.

The most common form takes in AC power from the mains supply and delivers a DC voltage to the item requiring power.

Accordingly power supplies are widely used in a variety of forms - some large supplying high levels of current, other power supplies, much smaller providing lower levels of power.

Power supply basics

The aim of a DC power supply is to provide the required level of DC power to the load using an AC supply at the input. Different applications require different attributes, but more often than not these days DC power supplies provide an accurate output voltage - this is regulated using electronic circuitry so that it provides a constant output voltage over a wide range of output loads.

In most power supplies there are number of different elements. These may not all be present in every design.

- ***Input transformer:*** The input transformer is used to transform the incoming line voltage down to the required level for the power supply. Typically the input transformer provides a step down function. It also isolates the output circuit from the line supply.
- ***Rectifier:*** The power supply rectifier converts the incoming signal from an AC format into raw DC. Either half wave or more commonly full wave rectifiers may be used as they make use of both halves of the incoming AC signal.
- ***Smoothing:*** The raw DC from the rectifier is far from constant falling to zero when the AC waveform crossed the zero axis, and then rising to its peak. The addition of a reservoir capacitor here fills in the troughs in the waveform, enabling the next stage of the power supply to operate. Large value capacitors are normally used within this stage.
- ***Regulator:*** This stage of the power supply takes the smoothed voltage and uses a regulator circuit to provide a constant output virtually regardless of the output current and any minor fluctuations in the input level.

Power supply regulation

There are two basic forms of power supply used in electronics equipment:

- ***Unregulated:*** This form of power supply was the only type used many years ago.

It simply consisted of a rectifier section and this was followed by capacitor or capacitor and inductor smoothing. There was no regulation to steady the voltage. If a large current was drawn the voltage would fall as a result of the resistive losses, and also the smoothing would not be as effective and the level of hum would rise.

- ***Voltage regulated:*** As transistor circuitry became more commonplace, regulated power supplies became more common. Today they are almost universally used. They typically incorporate a voltage reference, and the output voltage is compared to this and altered accordingly by control circuitry within the regulated power supply.

In addition to this, regulated power supplies may be further subdivided:

- ***Linear regulated power supply:*** Linear regulated power supplies use an analogue approach. A series element - a semiconductor transistor or FET - is controlled allow the correct voltage at the output for any current within the operating range

  Note on Linear Power Supplies:

  Linear power supplies are widely used for applications where low noise and ripple are required. As the name suggests, they use linear technology - typically a series linear regulator element to drop voltage. As such they dissipate power, but without any switching mode, they are able to offer high levels of performance

- ***Switching regulator power supply:*** The switching regulator format for a power supply uses a large output reservoir capacitor. A series element - a transistor or FET - is switched on and off to keep the voltage on the capacitor within the required limits.

  Note on Switch Mode Power Supplies:

  Switch mode power supplies and switch mode regulators have many advantages in terms of efficiency, size and weight. Their design can be more involved than might be thought at first. Yet with a good understanding, these switch mode power supplies, SMPSs, switch mode regulators and switch mode controllers can be successfully designed and built..

Each type of power supply regulation technique has its own advantages and disadvantages. As a result different types of regulator are used in different applications, although with technology improving, switching regulators are being used increasingly.

Regulator type	Advantages	Disadvantages
Linear regulator	• Very low level of noise • Straightforward technology	• Low level of efficiency • High levels of heat may need to be dissipated • Large size compared to switching regulator
Switching regulator	• Highly efficient • Can be made very compact • Low amounts of heat need to be removed	• Ripple and noise can be higher than linear regulator • EMC issues need to be addressed as switching spikes can cause interference

As a result of the different properties of each type of power supply regulator, linear regulators tend to be used in applications where very low levels of noise and ripple are required and heat dissipation may not be such a problem. Hi-fi amplifiers are one such area. Switching mode regulators are used more widely as they can be made very compact, they are very efficient and the levels of ripple, spikes and noise can normally be low enough for most applications.

APPLICATIONS OF TRANSISTOR – ITS USES

Applications of Transistors

The transistor as an amplifier

1. A transistor can be used to amplify current. This is because a small change in base current causes a large change in collector current.
2. Example is a microphone.
3. Sound waves that are fed into the microphone cause the diaphragm in the microphone to vibrate.
4. The electrical output of the microphone changes according to the sound waves.
5. As a result, the base current is varying because of the small alternating voltage produced by the microphone.
6. A small change in the base current causes a large change in the collector current.
7. The varying collector current flows into the loudspeaker. There, it is changed into the sound waves corresponding to the original sound waves.
8. The frequencies of both waves are equivalent but the amplitude of the sound wave from the loudspeaker is higher than the sound waves fed into the microphone.

- ***Component:*** Function
- ***Microphone:*** To change sound signal to electrical signal
- ***Capacitor:*** To block a steady current from flowing into the transistor and microphone.
- ***Potential divider:*** To apply a proportion of the total voltage across the emitter-base junction so that the junction is forward-biased.
- ***Transistor:*** To amplify the input wave form.
- ***Loudspeaker:*** To change the electrical signal to sound wave.

The transistor as switch

1. In a transistor, no current can flow in the collector circuit unless a current flows in the base circuit. This property allows a transistor to be used as switch.
2. The transistor can be turned on or off by changing the base.
3. There are a few types of switching circuits operated by transistors.

(a) ***Light-Operated Switch***

1. The circuit is designed to light the bulb in a bright environment and to turn it off in the dark.
2. One of the components in the potential divider is a light-dependent resistor (LDR). When it is placed in DARKNESS, its resistance is large. The transistor is switched OFF.
3. When LDR is lighted by bright light, its resistance falls to small value resulting in more supply voltage and raising the base current. The transistor is switched on, collector current flows and bulb lights up.

(b) ***Heat-operated switch***

1. One important component in the circuit of a heat-operated switch is the thermistor.
2. Thermistor is type of resistor that responds to the surrounding temperature. Its resistance increases when the temperature is low and vice versa.
3. When heat is applied to the thermistor, its resistance drops and a greater share of supply voltage is dropped across R. The base current increases followed by a greater increase in the collector current. The bulb will glow and the siren will sound.
4. This particular circuit is suitable as a fire alarm system.

Integrated Circuits (IC)

1. An integrated circuit (IC) consists of transistors, resistors, diodes and capacitors combined together in one wafer-thin chip of silicon.
2. This is one wafer-thin chip is called a microchip.

3. The microchip is only a few millimeters square with a thickness of 0.5 mm.

Advantages of an IC:

a. Consumes a small amount of electrical energy.
b. Very little heat is generated.
c. Occupies a small space which reduces the size of circuits.
d. Can be built at low cost.

MOSFET- FETs (Field Effect Transistors) of all types are widely used electronics components today. Of all the types of FET, the MOSFET (Metal Oxide Field Effect Transistors) is possibly the most widely used. MOSFETs provide many advantages. In particular they offer a very high input impedance and they are able to be used in very low current circuits. This is particularly important for integrated circuit technology where power limitations are a major consideration.

Figure 7.20: Symbols of of N Channel and P Channel MOSFETs. G-Gate, D-Drain and S-Source

MOSFET – Precautions when Handling

MOSFET Safety Precautions

Certain safety precautions must be observed when handling and using MOSFETs. It is important to check the manufacturer's specification sheet for maximum rating of EGS.(Voltage between Gate and Source.)

Caution!

If EGS is increased too much, the thin insulating layer ruptures, ruining the device. The insulating layer is so sensitive that it can be damaged by a static charge that has built up on the leads of the device. Electrostatic charges on fingers can be transferred to the MOSFET's leads when handling or mounting the device.

To avoid damage to the device, MOSFETs are usually shipped with the leads shorted together. Shorting techniques include wrapping leads with a shorting wire, inserting the device into a shorting ring, pressing the device into conductive foam, taping several devices together, shipping in antistatic tubes, and wrapping the device in metal foil.

Newer MOSFETs are protected with zener diodes electrically connected between the gate and source internally. The diodes protect against static discharges and in-circuit transients and eliminate the need for external shorting devices. In electronics, a transient is a temporary component of current existing in a circuit during adjustment to a load change, voltage source difference, or line impulse.

If the following procedures are followed, unprotected MOSFETs can be handled safely:

1. Prior to installation into a circuit, the leads should be kept shorted together.
2. The hand used to handle the device should be grounded with a metallic wristband.
3. The soldering iron tip should be grounded.
4. A MOSFET should never be inserted or removed from its circuit when the power is on.

QUESTIONS

1. What is the reason that MOSFETs have to be handled very carefully?
2. What voltage source, if exceeded, can ruin a MOSFET?
3. What methods are used to protect MOSFETs during shipping?
4. What precautions have been taken to protect newer MOSFETs?
5. Describe the procedures that must be observed when handling unprotected MOSFETs.

DIAC, SCR, TRIAC – Application

Diac

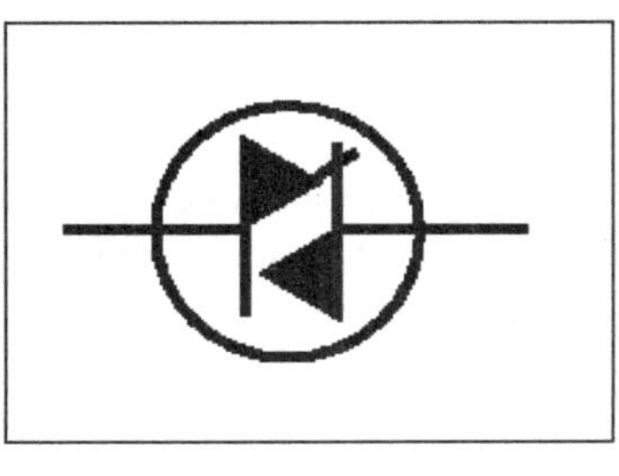

A diac is a form of solid-state switch used to switch AC voltage; it belongs to the class of switches known as thyristers. It is like a junction transistor without a base lead (it is a two-lead device) and accomplishes its switching action by breakdown at a certain voltage. There are also four layer devices with a similar mode of operation known as four-layer diodes. The DIAC is a full-wave or bi-directional semiconductor switch that can be turned on in both forward and reverse polarities.

The DIAC gains its name from the contraction of the words DIode Alternating Current.

The DIAC is widely used to assist even triggering of a TRIAC when used in AC switches. DIACs are mainly used in dimmer applications and also in starter circuits for florescent lamps.

Triac

The triode AC switch (TRIAC) is a power-switching device as is the SCR. The TRIAC conducts currents in both directions while the SCR allows current in only one direction. A common application is for lighting controllers. In response to a trigger, the triac conducts until the AC voltage applied reaches zero, then blocks flow until the next trigger occurs. Since a trigger can cause it to trigger current in either direction, it is an efficient power controller from essentially zero to full power. TRIACs are used in a number of applications. However they tend not to be used in high power switching applications - one of the reasons for this is the non-symmetrical switching characteristics. For high power applications this creates a number of difficulties, especially with electromagnetic interference.

However TRIACs are still used for many electrical switching applications:

- Domestic light dimmers
- Electric fan speed controls
- Small motor controls
- Control of small AC powered domestic appliances

The TRIAC is easy to use and provides cost advantages over the use of two thyristors for many low power applications. Where higher powers are needed, two thyristors placed in "anti-parallel" are almost always used.

The TRIAC is an electronic component that is widely used in many circuit applications, ranging from light dimmers through to various forms of AC control. It is generally only used for lower power applications, thyristors generally being used for the high power switching circuits.

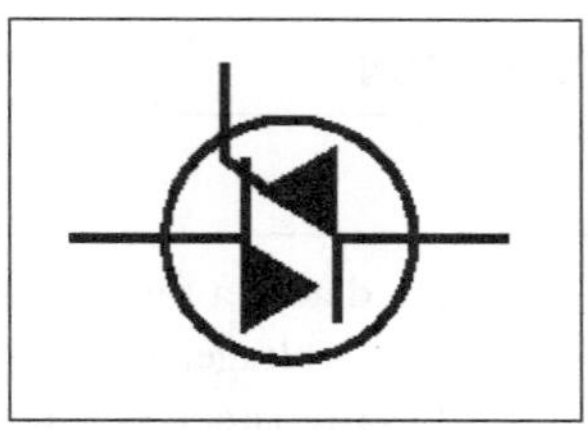

Symbol of TRIAC

Silicon Control Rectifier

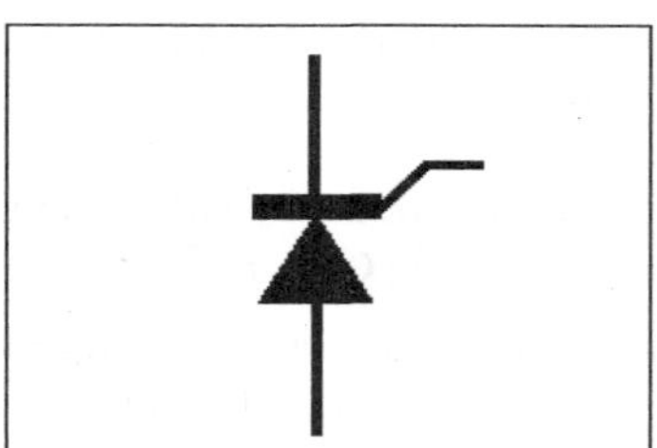

The SCR is a power-switching device commonly used for lighting control, motor speed control and other variable power applications.

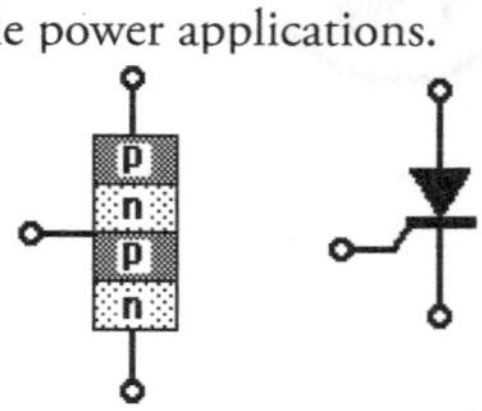

The silicon-controlled rectifier is like a junction transistor with a fourth layer and therefore three p-n junctions. The two outer junctions are forward biased by the voltage as shown, but the inner junction is reverse biased. A small current in the gate electrode can turn on the current, and it will stay on until the driving voltage is removed. It is called a rectifier because it conducts current in only direction. If AC voltage is applied, then it can be turned on by a pulse and remain on until the end of that half cycle. Timed 60 Hz triggers can by used to control power by changing the trigger point within the half cycle.

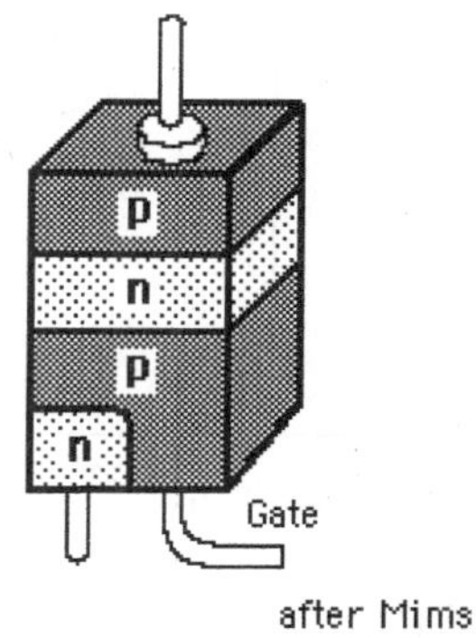

APPLICATIONS OF THYRISTOR

Thyristors are mainly used in devices where the control of high power, possibly coupled with high voltage is demanded. Their operation makes them suitable for use in medium to high-voltage AC power control applications, for example lamp dimming, controllers and motor control.

Practical: Identify the Fixed Voltage Regulators

Fixed Voltage Regulators

These regulators provide a constant output voltage. A popular example is the 7805 IC which provides a constant 5 volts output. A fixed voltage regulator can be a positive voltage regulator or a negative voltage regulator. A positive voltage regulator provides with constant positive output voltage. All those IC's in the 78XX series are fixed positive voltage regulators. In the IC nomenclature – 78XX ; the part XX denotes the regulated output voltage the IC is designed for. Examples:- 7805, 7806, 7809 etc. The regulated Output in 7805, 7806 AND 7809 is 5 V, 6 V and 9V respectively.

A negative fixed voltage regulator is same as the positive fixed voltage regulator in design, construction & operation. The only difference is in the polarity of output voltages. These IC's are designed to provide a negative output voltage. Example:- 7905, 7906 and all those IC's in the 79XX series. The regulated Output in 7905, 7906 AND 7909 is - 5 V,- 6 V and - 9V respectively.

Identify the pin diagram of Voltage regulator ICs

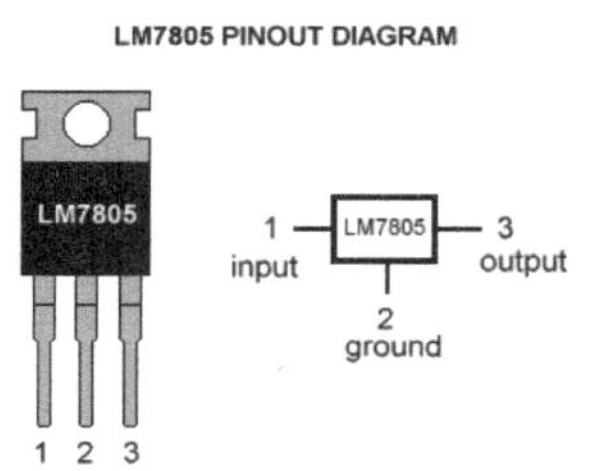

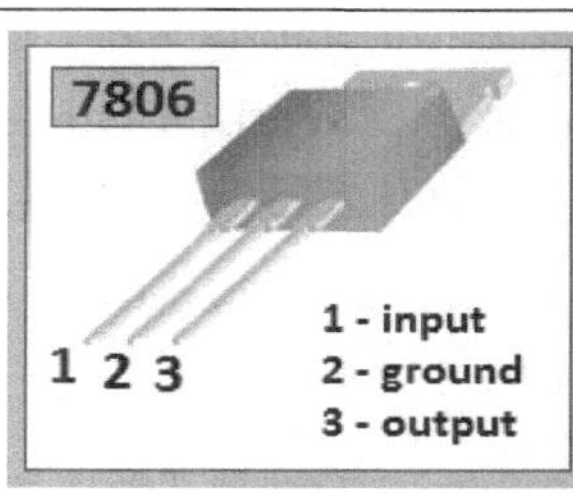

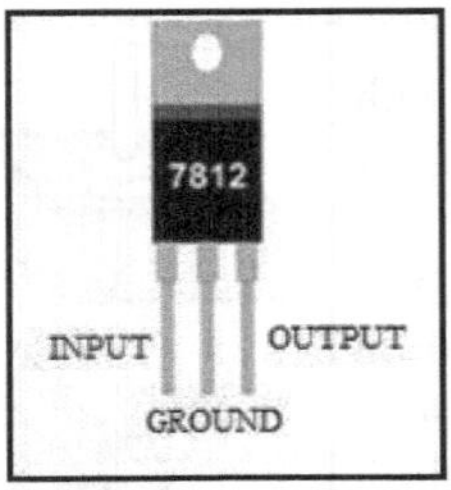	If you construct a Circuit with 7805/7806 and 7812 and measure the output between 2 and 3 it will be 5V/6V and 12 Volts respectively.
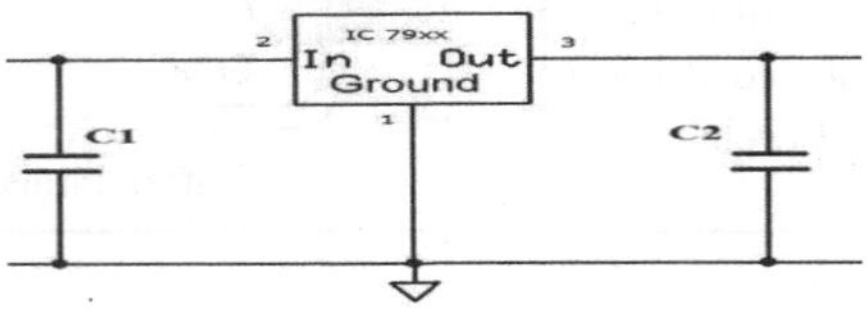	If you construct a Circuit with 7905/7906 and 7912 and measure the output between 2 and 3 it will be -5V/-6V and -12 Volts respectively.

Practical: Construct A 5 V Regulator Circuit with 7805

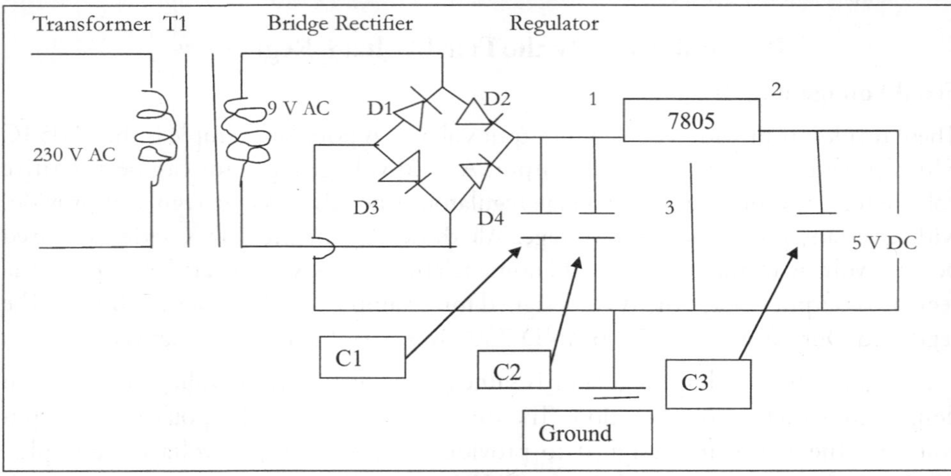

Figure 7.21: 5V Regulator Circuit

Parts List for a 5 V Regulator Circuit

1. T-1Step down Transformer- 230 V (Primary), 9V)Secondary -2 A
2. Bridge rectifier is with 4 Diodes- 1N 4007.
3. Capacitors C1= 470 μF-Electrolytic
4. Capacitor C2 and C3= .01μF Ceramic.
5. IC 7805= 5 V Regulator IC

Construction and working

Assemble the circuit either on a bread board or a PCB. Make the connections by soldering if using PCB. PCB is better as soldering practise can't be done on the Bread board.

Working

The Step down Transformer converts 230 V Mains AC to 9 V AC from the transformer. The Bridge circuit formed by FOUR DIODES act as a FULL WAVE RECTIFIER and gives and unregulated output slightly less than 9 V which has lot of ripples. The C1 acts as a flter and and removes the ripples and cleans the DC Voltage but it is fluctuating as the source mains fluctuates. The IC 7805 gives 5V DC regulated OUTPUT.

Testing- Measure the Voltages and record them at the following points

- At the input Primary Voltage=
- At the out put of the Transformer Secondary Voltage=
- At the output of the bridge rectifier=
- At the Output of the 7805

Practical: Construct a Variable Voltage Regulator using IC 723 and observe the O/P of Regulator output by varying the Input Voltage

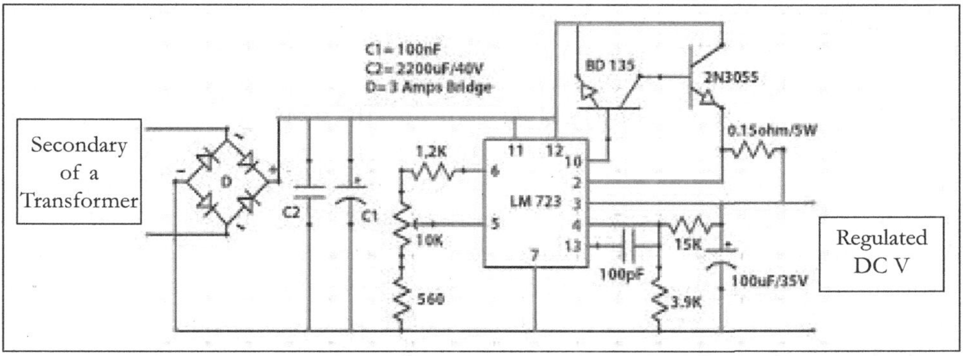

Figure 7.22: Variable Power Supply with 723

Assemble the circuit as shown above.

Put Heat Sinks on Transistor 2N 3055 . This is a Power Transistor to handle large currents. OF 2 A.

Voltage can be adjusted by Potentiometer of 10 K across Pin 5&6 of IC 723..

The Bridge Rectifier should be connected to 3 A Transformer. Can make the bridge rectifier by using 4 Diodes of 1N5402.

LCR METER

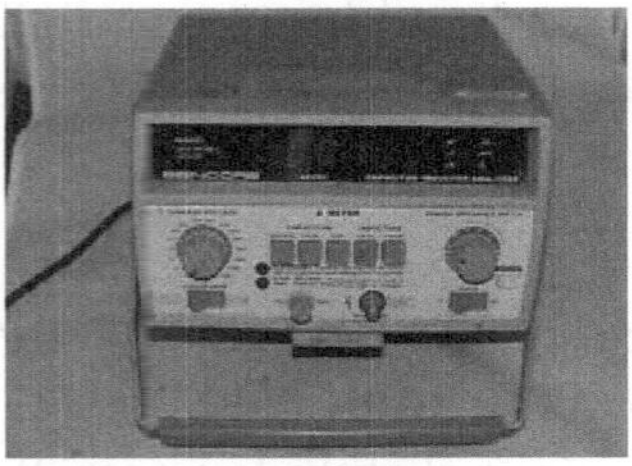

LCR-meter

An LCR meter is a piece of electronic test equipment used to measure the inductance (L), capacitance (C), and resistance (R) of a component. In the simpler versions of this instrument the true values of these quantities are not measured; rather the impedance is measured internally and converted for display to the corresponding capacitance or inductance value. Readings will be reasonably accurate if the capacitor or inductor device under test does not have a significant resistive component of impedance. More advanced designs measure true inductance or capacitance, and also the equivalent series resistance of capacitors and the Q factor of inductive components.

Usually the device under test (DUT) is subjected to an AC voltage source. The meter measures the voltage across and the current through the DUT. From the ratio of these the meter can determine the magnitude of the impedance. The phase angle between the voltage and current is also measured in more advanced instruments; in combination with the impedance, the equivalent capacitance or inductance, and resistance, of the DUT can be calculated and displayed. The meter must assume either a parallel or a series model for these two elements. The most useful assumption, and the one usually adopted, is that LR measurements have the elements in series (as would be encountered in an inductor coil) and that CR measurements have the elements in parallel (as would be encountered in measuring a capacitor with a leaky dielectric). An LCR meter can also be used to judge the inductance variation with respect to the rotor position in permanent magnet machines (however care must be taken as some LCR meters can be damaged by the generated EMF produced by turning the rotor of a permanent-magnet motor).

Hand held LCR meters typically have selectable test frequencies of 100 Hz, 120 Hz, 1 kHz, 10 kHz, and 100 kHz for top end meters. The display resolution and measurement range capability will typically change with test frequency.

Benchtop LCR meters typically have selectable test frequencies of more than 100 kHz. They often include possibilities to superimpose a DC voltage or current on the AC measuring signal. Lower end meters offer the possibility to externally supply these DC voltages or currents while higher end devices can supply them internally. In addition benchtop meters allow the usage of special fixtures to measure SMD components, air-core coils or transformers.

Bridge circuits

Inductance, capacitance, resistance, and dissipation factor can also be measured by various bridge circuits. They involve adjusting variable calibrated elements until the signal at a detector becomes null, rather than measuring impedance and phase angle.

Early commercial LCR bridges used a variety of techniques involving the matching or "nulling" of two signals derived from a single source. The first signal was generated by applying the test signal to the unknown and the second signal was generated by using a combination of known-value R and C standards. The signals were summed through a detector (normally a panel meter with or without some level of amplification). When zero current was noted by changing the value of the standards and looking for a "null" in the panel meter, it could be assumed that the current magnitude through the unknown was equal to that of the standard and that the phase was exactly the reverse (180 degrees apart). The combination of standards selected could be arranged to read out C and DF directly which was the precise value of the unknown standard. An example of this is the GenRad/IET Labs Model 1620 and 1621 Capacitance Bridges.

- ***Bridge method:*** This method uses a bridge which is balanced. At the balance point the bridge component positions can be used to determine the value of the component under test. This method is typically used for lower frequency measurements - often measurement frequencies of up to 100 kHz or so are used

  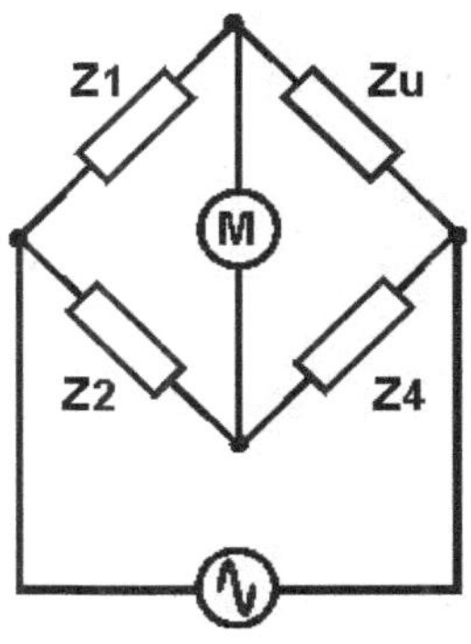

  Basic bridge based LCR meter configuration

  In bridge method uses a Wheatstone bridge configuration in which the device under test, DUT, is placed in a bridge circuit

  The DUT impedance is represented by Zu in the circuit. The impedance Z2 and Z3 are known. The oscillator circuit generally operates at frequencies up to about 100 kHz and can usually be selected before the test.

  Then Z1 can be changed until no current flows through D. This is the balance position for the bridge. AT this point the four impedances in the circuit obey the equation:

$$Zu = \left(\frac{Z3}{Z2}\right) Z1$$

This basic bridge circuit is sometimes used on its own in very primitive LCR meters. Some very old instruments actually have the elements that are manually balanced. However technology has moved on and higher levels of integration coupled with operational amplifier circuitry enable accurate automated versions of the circuit to be used.

CHAPTER

8

Inverters and UPS

Various types of batteries used in UPS and Inverters and their maintenance. Different types of inverter, UPS, Working principle, specifications, explanation with the help of block diagram.

Function of an Inverter

As the name suggests DC to AC inverter is an electronic device which is able to convert a DC potential normally derived from a lead-acid battery into a stepped-up AC potential which may be quite comparable to the voltage that is found in our domestic AC outlets.

An inverter normally may be divided into three important stages viz. oscillator, amplifier and the transformer output stage.

- ***Oscillator:*** This stage is basically responsible for the generation of oscillating pulses either through an IC circuit or a transistorized circuit. These oscillations are basically the productions of alternate battery positive and negative (ground) voltage peaks with a particular specified frequency (number of positive peaks per second.) Such oscillations are generally in the form of square pillars and are termed as square waves, and the inverters operating with such oscillators are called square wave inverters.

  The above generated square wave pulses though are too weak and can never be utilized to drive high current output transformers. Therefore these pulses are fed to the next amplifier stage for the required task.

- ***Booster or Amplifier (Driver):*** Here the received oscillating frequency is suitably amplified to high current levels using either power transistors or Mosfets. Though the boosted response is an AC, but is still at the battery supply voltage level and therefore cannot be used to operate electrical appliances which work at higher voltage AC potentials. The amplified voltage is therefore finally applied to the output transformer secondary winding.

- ***Output Transformer:*** We all know how a transformer works; in Ac/DC power supplies it is normally used to step-down the applied input mains AC to the lower specified AC levels through magnetic induction of its two windings. In inverters a transformer is used for similar purpose but with just opposite orientation, i.e. here the low level AC from the above discussed electronic stages is applied to the secondary windings resulting in an induced stepped up voltage across the primary winding of the transformer. This voltage is finally utilized for powering the various household electrical gadgets like lights, fans, mixers, soldering irons etc.

Diagram of Inverter

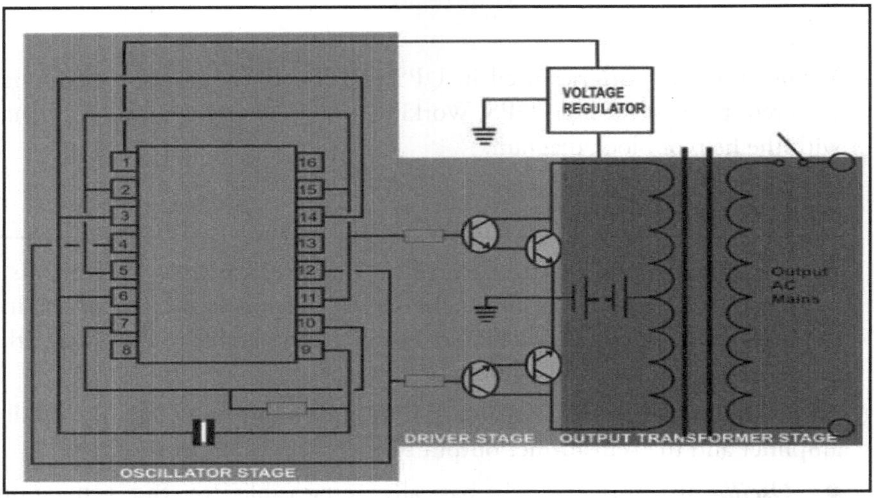

Figure 8.1: Inverter Block Diagram

DC to AC Inverter, Repairing Tips

In the above explanation a couple of things become very critical for obtaining correct results from an inverter. Firstly the generation of the oscillations, due to which the process of voltage induction is able to take place across the windings of the transformer. The second important factor is the frequency of the oscillations, which is fixed as per the country's specifications, for example countries that supply 230 V, generally have a working frequency of 50 Hz, in other countries where 120 V is specified mostly work at 60 Hz frequency.

Sophisticated electronic gadgets like TV sets, DVD players, computers etc. are never recommended to be operated with square wave inverters. The sharp rise and fall of the square waves are just not suitable for such applications.

However there are ways through more complex electronic circuits for modifying the square waves so that they become more favorable with the above discussed electronic equipment. Inverters using further complex circuits are able to produce waveforms almost identical to the waveforms available at our domestic mains AC outlets.

How to Repair an Inverter

Inverter is "Dead"- Check battery voltage and connection, check for a blown fuse. If that's OK, open the inverter outer cover and perform the following:

Locate the oscillator section; disconnect its output from its preceding stage and using a frequency meter confirm its proper working. No frequency or a stable DC indicates a possible fault with the stage. Check its IC and the associated components for the remedy.

In case you find the oscillator stage working fine, go for the next stage i.e. the amplifier stage. Check each device using a digital multimeter, you may have to completely remove them from the board for the particular tests. If you find a particular device to be faulty just replace it with a new one.

Sometimes transformers also become the major cause for a malfunction. Check for an open winding or a loose internal connection in the associated transformer. If you find it to be suspicious, immediately change it with a new one.

Maintenance of Inverter Batteries

Inverter batteries are called secondary batteries as these are rechargeable.

Inspection

There are many tools that may help in caring for and maintaining batteries

Recommended Spares:

Wrench	Voltmeter	Hydrometer	Post Cleaner
Vaseline	Goggles	Gloves	Baking Soda

Safety Tips

Always USE PERSONAL protective clothing, gloves, and goggles when handling batteries.

Batteries should be carefully inspected on a regular basis in order to detect and correct potential problems before they can do harm. It is a great idea to start this routine when the batteries are first received.

Inspection Guidelines:

1. Examine the outside appearance of the battery
   - Look for cracks in the container.
   - The top of the battery, posts, and connections should be clean, free of dirt, fluids, and corrosion. If batteries are dirty, refer to the cleaning section for the proper cleaning procedure.
   - Repair or replace any damaged batteries.

2. Any fluids on or around the battery may be an indication that electrolyte is spilling, leaching, or leaking out.
   - ❖ Leaking batteries must be repaired or replaced.
3. Check all battery cables and their connections.
   - ❖ Look closely for loose or damaged parts.
   - ❖ Battery cables should be intact; broken or frayed cables can be extremely dangerous.
   - ❖ Replace any cable that looks suspicious.
4. Tighten all wiring connections. Make certain there is good contact with the terminals.

Cleaning

Batteries seem to attract dust, dirt, and grime. Keeping them clean will help one spot trouble signs if they appear and avoid problems associated with grime.

1. Check that all vent caps are tightly in place.
2. Clean the battery top with a cloth or brush and a solution of baking soda and water.
   - ❖ When cleaning, do not allow any cleaning solution, or other foreign matter to get inside the battery.
3. Rinse with water and dry with a clean cloth.
4. Clean battery terminals and the inside of cable clamps using a post and clamp cleaner.
   - ❖ Clean terminals will have a bright metallic shine.
5. Reconnect the clamps to the terminals and thinly coat them with petroleum jelly (Vaseline) to prevent corrosion.
6. Keep the area around the batteries clean and dry.

Battery Safety Precautions

Lead-acid batteries contain sulphuric acid, which can cause severe burns besides generating hydrogen and oxygen during charging, the mixture of which can be explosive.

To help avoid risk of danger and injury, observe these precautions when handling or working with a lead-acid battery .

- Always store or recharge lead acid batteries in a well ventilated area away from sparks or open flames.
- When recharging or handling lead acid batteries, wear acid resistant goggles/face shield, gloves, and if available, an apron.

- Always keep lead acid battery vent caps securely in place. However, while charging, it should be kept open.
- Never overcharge a lead acid battery and replenish electrolyte with battery grade distilled/ demineralised water only.
- Lead acid storage and charging areas should be posted with "Flammable" "No Smoking" signs.
- If acid gets into your eye(s), flush immediately with water for 15 minutes and then promptly seek medical attention.
- Emergency eye wash stations should be located near lead acid battery storage and charging areas.
- If acid gets on your skin, rinse the affected area immediately with large amounts of water. Seek medical attention if the chemical burn appears to be second degree or higher.

CHAPTER 9

Switch Mode Power Supply

BLOCK DIAGRAM OF SWITCH MODE POWER SUPPLIES AND THEIR WORKING PRINCIPLES

Power supply is an electronic circuit that is used for providing the electrical power to appliances or loads such as computers, machines, and so on. These electrical and electronic loads require various forms of power at different ranges and with different characteristics. So, for this reason the power is converted into the required forms (with desired qualities) by using some power electronic converters or power converters.

Electrical and electronic loads work with various forms of power supplies, such as AC power supply, AC- to-DC power supply, High-voltage power supply, Programmable power supply, Uninterruptable power supply and Switch-mode power supply.

WHAT IS SWITCH-MODE POWER SUPPLY?

The electronic power supply integrated with the switching regulator for converting the electrical power efficiently from one form to another form with desired characteristics is called as Switch-mode power supply. It is used to obtain regulated DC output voltage from unregulated AC or DC input voltage.

Switch mode power supply

Similar to other power supplies, switch-mode power supply is a complicated circuit that supplies power from a source to loads. switch-mode power supply is essential for power consuming electrical and electronic appliances and even for building electrical and electronic projects.

Types of Switch Mode Power Supply

There are different types of SMPS, among those, a few are as follows

- DC to DC converter
- AC to DC converter
- Fly back converter
- Forward converter

Switch Mode Power Supply's Working Principle

The working of a few types of switch-mode power supply is as follows:

1. DC to DC Converter SMPS Working Principle: In a DC-to-DC converter, primarily a high-voltage DC power is directly obtained from a DC power source. Then, this high-voltage DC power is switched at a very high switching speed usually in the range of 15 KHz to 50 KHz.

   And then it is fed to a step-down transformer which is comparable to the weight and size characteristics of a transformer unit of 50Hz. The output of the step-down transformer is further fed into the rectifier. This filtered and rectified output DC power is used as a source for loads, and a sample of this output power is used as a feedback for controlling the output voltage.With this feedback voltage, the ON time of the oscillator is controlled, and a closed-loop regulator is formed.

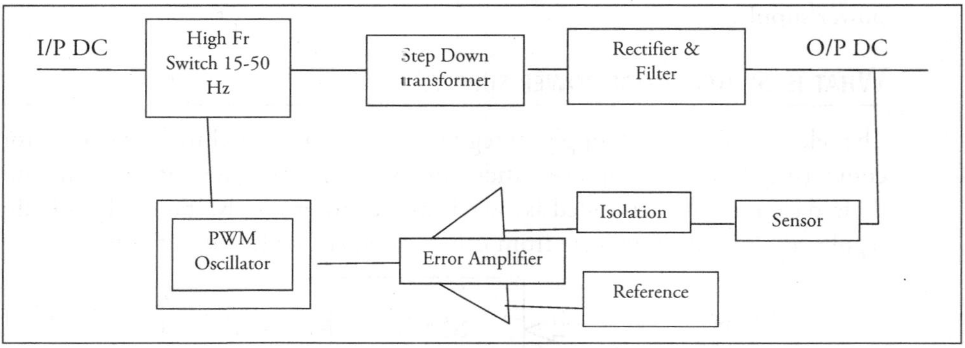

Figure 9.1: DC to DC converter SMPS

The output of the switching-power supply is regulated by using PWM (Pulse Width Modulation). As shown in the circuit above, the switch is driven by the PWM oscillator, such that the power fed to the step-down transformer is controlled indirectly, and hence, the output is controlled by the PWM, as this pulse width signal and the output voltage are inversely proportional to each other.

If the duty cycle is 50%, then the maximum amount of power is transferred through the step-down transformer, and, if duty cycle decreases, then the amount of power transferred will decrease by decreasing the power dissipation.

2. ***AC to DC Converter SMPS Working Principle:*** The AC to DC converter SMPS has an AC input. It is converted into DC by rectification process using a rectifier and filter. This unregulated DC voltage is fed to the large-filter capacitor or PFC (Power Factor Correction) circuits for correction of power factor as it is affected. This is because around voltage peaks, the rectifier draws short current pulses having significantly high-frequency energy which affects the power factor to reduce.

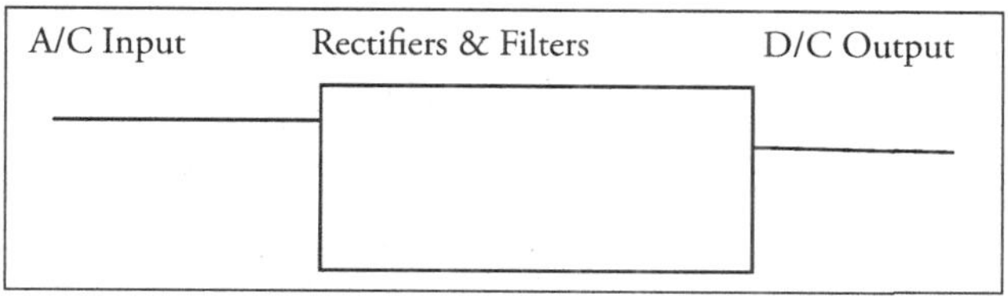

Figure 9.2: AC to DC converter SMPS

The rest of the circuit will be same. The A/C is converted to DC and like the DC-DC Converter it is fed to the DC input.

It is almost similar to the above discussed DC to DC converter, but instead of direct DC power supply, here AC input is used. So, the combination of the rectifier and filter, shown in the block diagram is used for converting the AC into DC and switching is done by using a power MOSFET amplifier with which very high gain can be achieved. The MOSFET transistor has low on-resistance and can withstand high currents. The switching frequency is chosen such that it must be kept inaudible to normal human beings (mostly above 20KHz) and switching action is controlled by a feedback utilizing the PWM oscillator.

This AC voltage is again fed to the output transformer shown in the figure to step down or step up the voltage levels. Then, the output of this transformer is rectified and smoothed by using the output rectifier and filter. A feedback circuit is used to control the output voltage by comparing it with the reference voltage.

3. ***Fly-back Converter type SMPS Working Principle:*** The SMPS circuit with very low output power of less than 100W (watts) is usually of Fly-back converter type SMPS, and it is very simple and low- cost circuit compared to other SMPS circuits. Hence, it is frequently used for low-power applications.

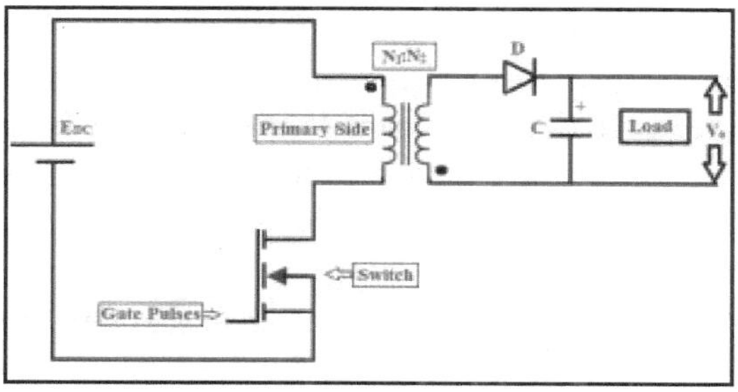

Figure 9.3: Fly-back Converter type SMPS

The unregulated input voltage with a constant magnitude is converted into a desired output voltage by fast switching using a MOSFET; the switching frequency is around 100 kHz. The isolation of voltage can be achieved by using a transformer. The switch operation can be controlled by using a PWM control while implementing a practical fly-back converter.

Fly-back transformer exhibits different characteristics compared to general transformer. The two windings of the fly-back transformer act as magnetically coupled inductors. The output of this transformer is passed through a diode and a capacitor for rectification and filtering. As shown in the figure, the voltage across this filter capacitor is taken as the output voltage of the SMPS.

4. ***Forward Converter type SMPS Working:*** Forward converter type SMPS is almost similar to the Fly-back converter type SMPS, but in the forward converter type, a control is connected for controlling the switch and at the output of the secondary winding of the transformer, and the rectification and filtering circuit is complicated as compared to the fly-back converter.

   It can be called as a DC to DC buck converter, along with a transformer used for isolation and scaling. In addition to the diode D1 and capacitor C, a diode D2 and an inductor L are connected at the output end. If switch S gets switched ON, then the input is given to the primary winding of the transformer, and hence, a scaled voltage is generated at the secondary winding of the transformer.

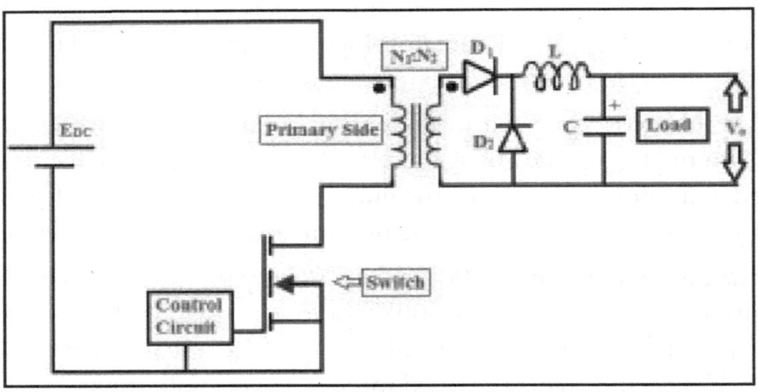

Figure 9.4: Forward Converter type SMPS

Thus, the diode D1 gets forward biased and scaled voltage is passed through the low-pass filter preceding the load. If the switch S is turned off, then the currents through the primary and secondary winding reach to zero, but the current through the inductive filter and load can not be change abruptly, and a path is provided to this current by the freewheeling diode D2. By using the filter inductor, the required voltage across the diode D2 and to maintain the EMF required for maintaining the continuity of the current at inductive filter.

Even though the current is diminishing against the output voltage, approximately the constant output voltage is maintained with the presence of the large capacitive filter. It is frequently used for switching applications with a power in the range of 100 W to 200 W.

Different types of topologies are there in which SMPS can be realized such as Buck converter, Boost converter, Self Oscillating fly-back converter, Buck-boost converter, Boost-buck, Cuk, Sepic. But only a few are discussed in this article, namely DC to DC converter, AC to DC converter, Fly-back converter and Forward converter. For more information regarding the types of switch-mode power supply and the types of SMPS with their working principles, feel free to write your comments for improving this article technically so that you can help the other readers to get awareness of SMPS.

CHAPTER

10 Washing Machines

Washing M/c: different types of machines, washing techniques, parts of manual, semi automatic and fully automatic machines, basic working principle of manual, semi automatic and fully automatic machines, study the working of motors, different types of timers, power supply circuits

CLOTHES WASHING MACHINE

If there's one household appliance most of us simply could not do without, it's the clothes washer. If you've ever been without your machine for a few days or weeks, you'll know just how hard it is to wash clothes by hand. Although clothes washers look pretty straightforward, they pull off a really clever trick: with the help of detergents, they separate the dirt from your clothes and then rinse it away. But how exactly do they work?

Photo: A typical clothes washing machine, powered by electricity. This one is a front-loader: you put your clothes into that little circular window at the front. In the United States and Asia, top-loading machines are more common.

THE PARTS OF A CLOTHES WASHER

Photo: Inside a clothes washer drum. The paddles turn the clothes through the water. The holes let the water in (from above) and out (from below). The rubber seal stops water leaking out through the door.

The basic idea of a clothes washer is simple: it revolves the clothes about soaked in soap powder called detergent for a while and then spins fast to remove the water afterward. But there's a bit more to it than that. Think of a clothes washer and you probably think of a big drum that fills with water—but there are actually two drums, one inside the other.

The inner drum is the one you can see when you open the door or the lid. In a front-loading clothes washer, the drum stands upright. You push your clothes inside the door from the front and the whole drum rotates about a horizontal axis (like a car wheel). The drum has lots of small holes to let water in and out and paddles around the edge to slosh the clothes around. In a top loader, more common in the United States and Asia, you open a lip on top and drop your clothes into the drum from above. The drum is mounted about a vertical axis but doesn't actually move. Instead, there's a paddle in the middle of it called an agitator that turns the clothes around in the water.

The two drums are the most important parts of a clothes washer, but there are lots of other interesting bits too. There's a thermostat (thermometer mechanism) to test the temperature of the incoming water and a heating element that warms it up to the required temperature. There's also an electrically operated pump that removes water from the drum when the wash is over. There's a mechanical or electronic control mechanism called a programmer, which makes the various parts of the clothes washer go through a series of steps to wash, rinse, and spin your clothes. There are two pipes that let clean hot and cold water into the machine and a third pipe that lets the dirty water out again. All these pipes have valves on them (like little doors across them that open and shut when necessary).

The Washing Machine Program

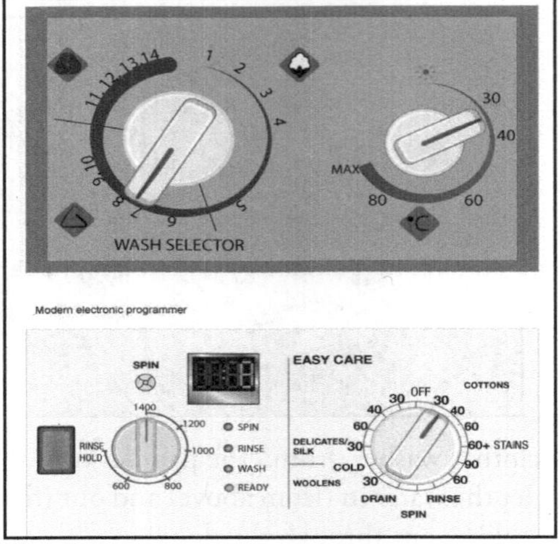

Photo: Controlling a washing machine: An old-style mechanical clothes washer programmer. The dial on the left selects the program. The dial on the right sets the wash temperature (it's effectively a thermostat). Bottom: A modern electronic programmer. These dials are mounted on a computerized programmer circuit. The countdown-display tells you how long in hours and minutes it will be before your washing is clean and ready to take out (one hour and two minutes in this case, for a 30°C wash with a very fast 1400rpm spin).

All the important parts of the clothes washer are electrically controlled, including the inner drum, the valves, the pump, and the heating element. The programmer is like the conductor of an orchestra, switching these things on and off in a sensible sequence that goes something like this:

1. You put your clothes in the machine and detergent either in the machine itself or in a tray up above.
2. You set the program you want and switch on the power.
3. The programmer opens the water valves so hot and cold water enter the machine and fill up the outer and inner drums. The water usually enters at the top and trickles down through the detergent tray, washing any soap there into the machine.
4. The programmer switches off the water valves.
5. The thermostat measures the temperature of the incoming water. If it's too cold, the programmer switches on the heating element. This works just like an electric kettle or water boiler.
6. When the water is hot enough, the programmer makes the inner drum rotate back and forth, sloshing the clothes through the soapy water.
7. The detergent pulls the dirt from your clothes and traps it in the water.
8. The programmer opens a valve so the water drains from both drums. Then it switches on the pump to help empty the water away.
9. The programmer opens the water valves again so clean water enters the drums.
10. The programmer makes the inner drum rotate back and forth so the clean water rinses the clothes. It empties both drums and repeats this process several times to get rid of all the soap.
11. When the clothes are rinsed, the programmer makes the inner drum rotate at really high speed—around 80 mph (130 km/h). The clothes are flung against the outside edge of the inner drum, but the water they contain is small enough to pass through the drum's tiny holes into the outer drum. Spinning gets your clothes dry using the same idea as a centrifuge.
12. The pump removes any remaining water from the outer drum and the wash cycle comes to an end.
13. You take your clothes out and marvel at how clean they are!
14. But there's still the problem of drying your wet clothes to figure out.

Why do washing machines need so many programs?

Your machine doesn't know what you put into it and can't automatically tell how carefully to wash something like a delicate woollen jumper—because it doesn't know that's what it's got to do! The only things under its control are the amount and temperature of the water, the speed of the spin, the number of times the drum oscillates, the number of rinses, and so on. No-one wants to wash clothes in a scientific way: "I think I need 5.42 litres of water at exactly 42°C, I'll need to wash for exactly 34 minutes, and I'll need 200 spin revolutions when I'm done." That would give us literally an infinite number of possibilities, which is too much like hard work. Recognizing this, machine engineers try to make life easy by offering a few preset programs: each one uses a slightly different combination of these variables so it washes safely within the tolerance of different fabrics.

Why does that matter? All fabrics are different. A fabric like wool is immensely strong but has two big drawbacks (from the point of getting it clean): it's extremely hygroscopic (absorbs huge amounts of water) and loses its elasticity as the temperature increases. So if you're designing a washing machine to wash woollens, that's your starting point: don't allow the wool to become too hot (because the fibers will degrade and stretch too much) and don't agitate it excessively because it will stretch and not return to shape. With sturdier fabrics like denim, you can afford to bash them about in the drum much more—indeed, you must do so, because you need the agitation to get the detergent deep into the fibers and break up the dirt (and, of course, clothes made from denim are more likely to get dirtier than more delicate fabrics such as cashmere jumpers, which people treat more carefully).

Each program you find marked on a clothes washing machine is a best guess by the engineers as to how much agitation a particular garment/fabric is likely to need and how much it can put up with without getting damaged. If you were handwashing in a sink, you'd make those judgements instinctively, balancing the need to get your garment clean with the need to protect it from damage. While your brain/hands would do that without thinking, the washing machine does it with a certain wash temperature, so many agitations, so many spins, and a certain spin speed.

But do machines really need so many programs?

Look at the programmers in the photos above and you'll see something interesting: both machines seem to have an incredible number of programs. The mechanical programmer in the top photo offers 14 programs, seven temperatures, two spin speeds, and full or half load—and if you multiply those you'll get 392 possibilities! The electronic programmer underneath it offers 12 programs, 5 spin speeds, and various other options so, again, a good few hundred possibilities. Yet if you're like me, you probably wash almost all your clothes on a single program all the time. Even if you don't do that, it's unlikely you could think of 392 different types of clothing that need washing in 392 different ways.

Much of this is a marketing con to make you believe the machine has more features than it really does. Most machines can really do only about three or four basic washes: 1) a high-temperature, long-duration wash for white laundry that uses a fairly high spin speed and lots of water; 2) a slightly faster, lower-temperature wash for colored cottons that uses similar spin speed and water volume; 3) a synthetics wash that uses the same amount of water, agitates the laundry less, spins more slowly, and uses lower temperatures; and 4) a woollens wash that probably uses quite a bit more water, but agitates the drum less, and spins the water out relatively slowly. Any other programs are variations of these four.

What's the difference between a front-loader and a top-loader?

The washing process is slightly different in front- and top-loading machines, so let's now look at each of these in a bit more detail:

In a front-loading clothes washer...

1. There's a fixed outer drum (blue) and a rotating inner drum (red) with small holes around its edge. The drums are mounted on a horizontal axis.
2. The inner drum is held to the frame of the machine by heavy-duty springs. That's because, when the clothes spin, they can make the drum shake violently; the springs help to absorb the vibrations.
3. Hot and cold water enter through the detergent tray at the top.
4. The inner drum turns back and forth. The plastic paddles on the inside (shown here by gray triangles) help to slosh the clothes through the detergent and water held by the outer drum.
5. An electric motor turns the inner drum, typically using a long rubber belt (yellow).
6. A heating element heats the water as necessary.
7. When the wash cycle is finished, the pump sucks the water away.
8. The water empties down a tube to the drain.

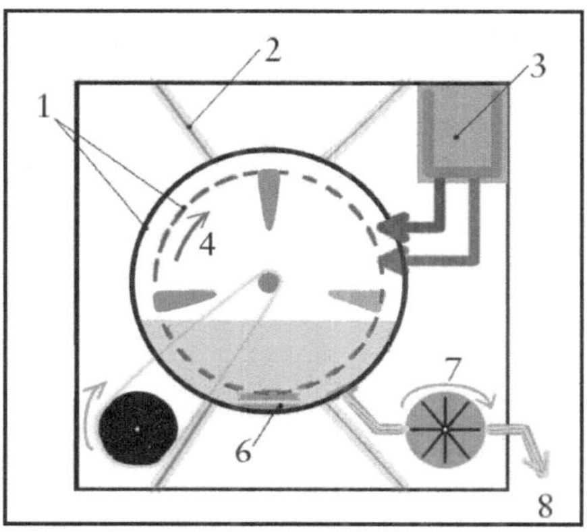

Figure 10.1: Front loader Washing

In a Top loader Washing Machine

1. Lift the lid on top and drop your clothes in from above. We're looking here from one side.
2. Just like in a front-loading machine, there's an outer drum (blue) and an inner drum with holes in it (red), but they're mounted about a vertical axis.
3. Hot and cold water enter through pipes near the top, passing through the detergent tray and flushing the detergent into the machine.
4. During the wash cycle, a large plastic agitator (green) turns around, moving your clothes through the water. Both drums remain stationary.
5. The agitator is powered by an electric motor using a rubber belt.
6. During the spin cycle, the same electric motor turns the inner drum (red) at high speed, throwing water through its holes into the outer drum.
7. When the wash is finished, the pump drains the water from the outer drum.

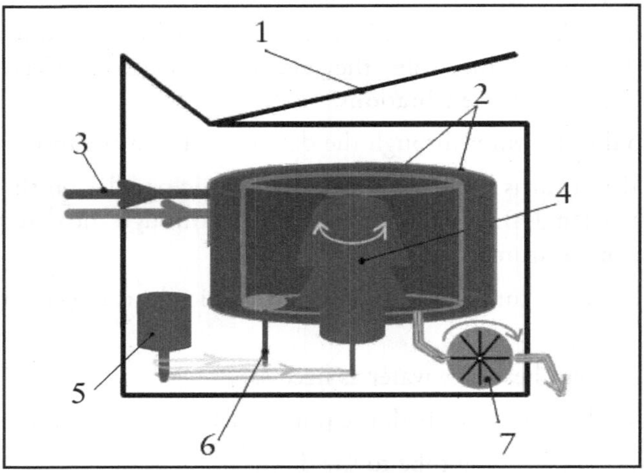

Figure 10.2: Top loader Washing Machine

CHAPTER

11

Microwave Oven

Microwave oven: Different types of oven, study the various functions of Oven, Electrical wiring diagram of microwave oven, working of Power supply.

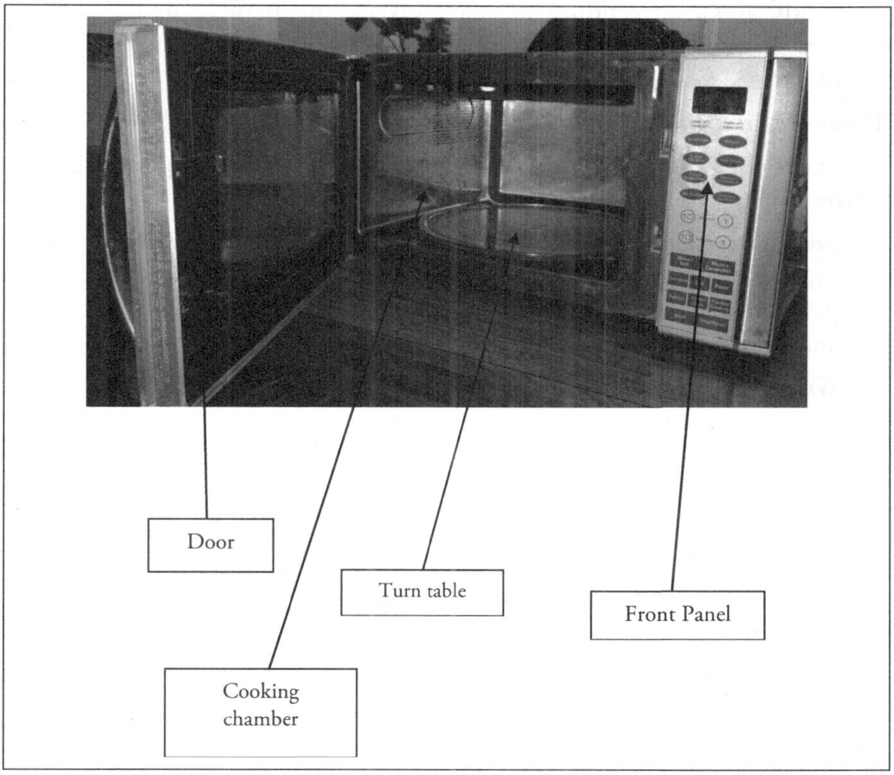

Figure 11.1: Parts of A microwave oven from the front

A Microwave Oven has an opening in the front fitted with a **door**. The microwave Oven can not be turned on with the door open. The door is fitted with a microswitch that will shut the Microwave automatically when the door is kept open and the unit is in switched on condition. The inside where the food is heated is made of steel as microwaves can't pass through steel.

Located beside the door there is a **control panel** and a **display section** . The control panel houses a **circuit board** which controls the operations and the functions of the Microwave Oven.

Magnetron is the device which generates Microwaves and is the heart of the Microwave Ovens.

There is a Cooling **fan inside which** pulls air through vents in the side and across the electronics board when the oven is on. The air is blown out of the fan and through the cooling fins of the magnetron.

A **turn table** rotates automatically with the help of a motor and rotates clockwise and anti-clock wise when cooking food. The food is cooked uniformly from all sides because of this rotation of the tern table

In Grill and Convection Microwave Ovens, **metal racks or trays** are provided to place the food items closer to the grill placed at the top. This results in uniform roasting of the food even at the top.

Types of Microwave Ovens:

Generally the microwave ovens are available in three categories. solo, grill & convection.

1. ***Solo Microwave Ovens:*** The solo models are basic models in micro wave ovens. These ovens only have Magnetrons to produce Microwaves and thus can perform the heating and boiling. However it can't perform roasting or grilling and baking operations.
2. ***Grill Microwave Ovens:*** The grill models have heating coils which enables grilling or roasting functions. It can be used with the Microwave ON as well as OFF position. For example we can use both these functions simultaneously while cooking chicken legs, fish or paneer dishes.
3. ***Convection Microwave Ovens:*** In the case of convection models the micro wave oven has a blower to bake the food. The heaters are connected to a thermostat to control the cavity temperature, the range of the thermostat is 95° c to 230° c. This is a very versatile function and very convenient for making cake, biscuits, naan-katai or even tandoori dishes. This can use all the functions in combination or independently.

How the Microwave Oven Functions

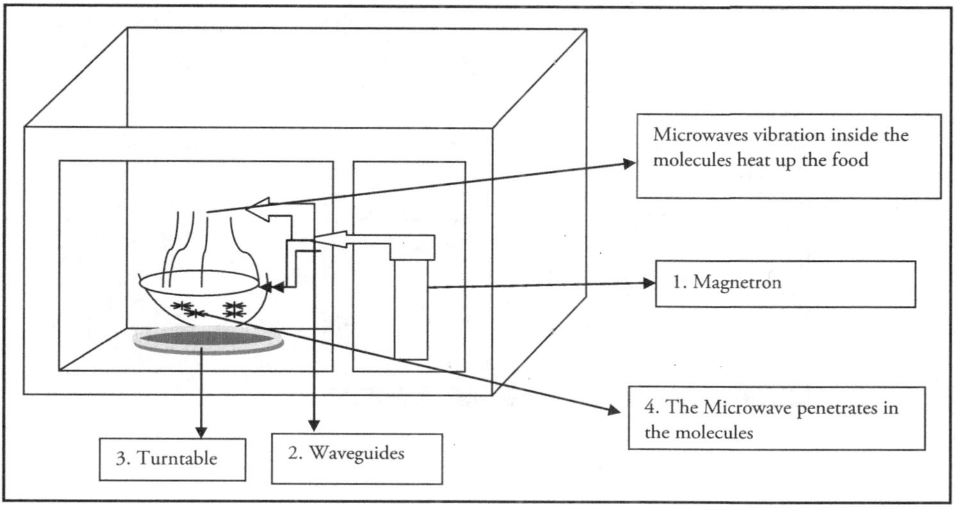

Figure 11.2: Microwave Functional Diagram

1. Microwave Generator also called the Magnetron is the heart of the Microwave oven.

   The frequency of Microwave generated is 2.54 GHz or $2.54x10^9$ Hz i.e the molecules of the microwave vibrate $2.54x10^9$ times in one second. The wavelength of this vibration is calculated by the WAVE EQUATION:

Velocity = Wavelength X Frequency OR Wavelength= Velocity / Frequency

   Velocity of light= $3x10^{10}$ cm per sec; Frequency= $2.54x10^9$ per sec.

   Thus the wavelength of Microwave is = ($3x10^{10}$ cm/sec) / ($2.54x10^9$ /sec)= 12 cm.

2. The magnetron forces these microwaves into the food compartment through a channel called a wave guide.

3. The food bowel lies on a turntable, spinning slowly round so the microwaves cooking takes place evenly.

4. The microwaves gets reflected from the metallic wall of the cavity and bounces back and forth. As they reach the food particles the waves penetrate the food particles and they make the molecules of the food particle to vibrate briskly as the frequency is very high.

5. Vibrating molecules have heat so, the faster the molecules vibrate, the hotter the food becomes. Thus the microwaves pass their energy onto the molecules in the food, rapidly heating it up.

Microwave important components identification after opening

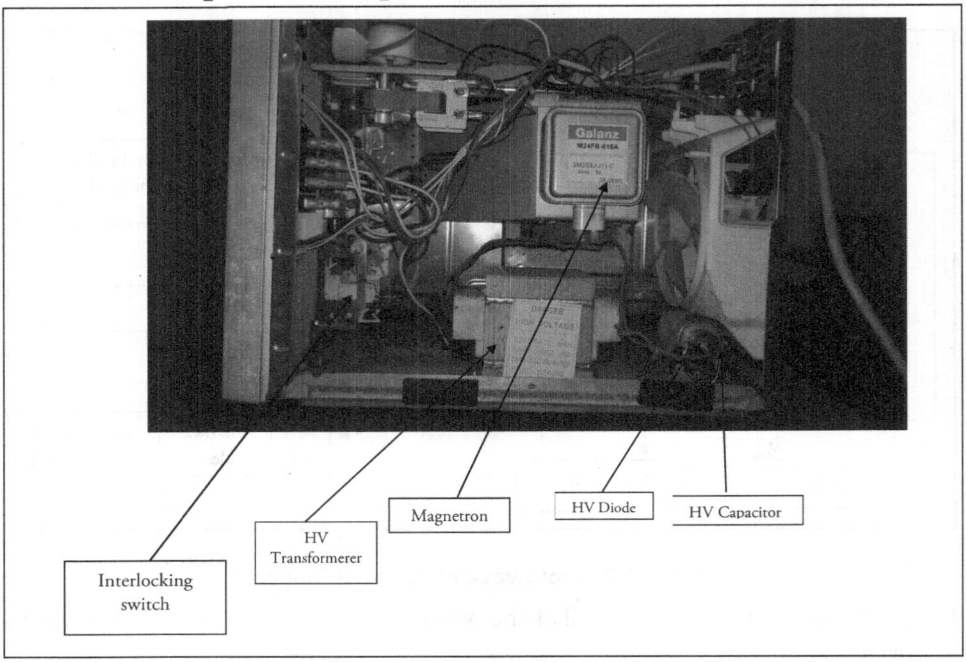

Figure 11.3: Important Components of Microwave Oven

The Microwave Front Panel Controls:

The figure below identifies the Front Panel controls of Microwave. The important controls are as below:

1. ***Start:*** This button will start the Microwave.
2. **Stop/reset:** This button will stop the cooking.
3. ***Various Modes:*** Like Convection, Grill, Defrost.
4. Timer control.
5. Various timing settings.

Troubleshooting Guide

Practical exercises- You can do the following important practical with this troubleshooting Guide:

1. Rectify the faults leading to Fuse Blow off when cooking is initiated. Check whether there is a short in the HV Capacitor.
2. Rectify the fault leading to non response of touch switches (front panel).
3. Rectify the fault leading to dead set.
4. Rectify the fault leading to long cook time.

Important Safety Information: Microwave ovens are among the most dangerous appliance to work on. It has High Voltages and can be fatal to work without safety gears. Adequate precautions are required while troubleshooting and repairing.

Problem-General	Specific Symptoms	Action to be taken
Heating or cooking problems	Oven runs but does not heat	Check if HV Diode is short. Check if HV Transformer is OK. Check if Magnetrom is short. Check if Magnetron dome (or antenna) damaged from arcing (if this is the case, check for a stalled stirrer/antenna). Check if there is a Magnetron Insulator breakdown. Have a visual inspection for Short in HV wiring. Short HV Capacitor- Check if Fuse blows after start of cooking- Check with HV Diode-whether it is Open. Test for Open Magnetron Filament Check if Intermittent or oxidized connection to one or both magnetron filament terminals. Check if the HV transformer is Open Circuit(usually the primary winding) Test if the interlock switch is defective Burned slip-on connector to interlock switch. Inspect and repair as necessary. Test for Defective triac. Test if there is Open fuse in HV circuit (more common in commercial models) Test if there is Open HV capacitor (uncommon)

	Low and slow heat	Measure the Line Voltage Magnetron will need replacement in the following cases: - Magnetron used for more than 2,000 hours. -Cracked magnet(s) can be fixed after inspection. -Burned dome (antenna) can be fixed after inspection -Magnetron insulator breakdown can be fixed after inspection
	Intermittent heat	Intermittent or oxidized/burned connection to one or both magnetron filament terminals_ can be fixed after inspection . There can be a burned slip-on connector due to a poor crimp joint, or weakened connection. Which can be repaired or replaced as necessary.
	Hot spots because of uneven heating	Check for mechanical bind, loose terminal connection or open motor windings
	Starts cooking by itself OR Only cooks at full power	Test for Defective triac and replace if required.
Mains Power Problem	Completely dead (as though unplugged) Goes dead intermittently Goes dead in mid-cycle (overheats)	Check for mechanical bind, loose terminal connection or open motor windings. Inspect and Clean as necessary Inspect and Clean as necessary
Timer, Control Panel or Function Problem	Timing problem	Check for mechanical bind, loose terminal connection or open motor windings
	Control Panel and mode selection or operations problem	Defective touch panel (keypad). NOTE: If none of the pads respond, the problem could be elsewhere.
	Display Problem	Defective control board (main printed circuit board) or sensor unit.) or Display board defective. Replace or repair as required

Function of Important Parts of a Microwave Oven

1. ***The High Voltage Rectifiers:*** The high-voltage rectifier (diode) works along with the high-voltage capacitor to effectively double the already-high voltage that is

provided by the power transformer. This powerful voltage, about 3000 - 5000 volts DC (depending on the model), is applied to the magnetron tube, causing it to produce the microwave energy that cooks the food.

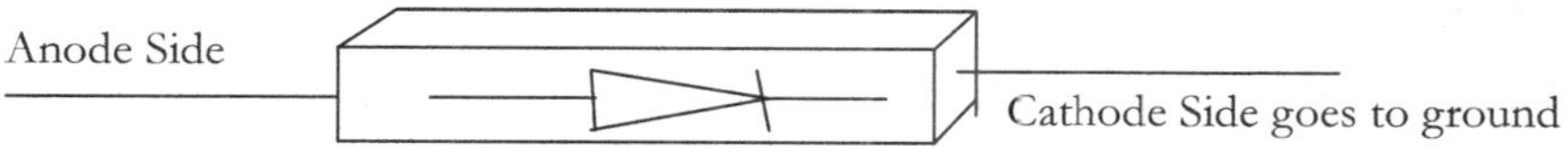

Caution- before doing any test unplug the microwave oven and discharge all hv capacitors.

Testing the HV diode requires an ohm meter with at least a 6 volt battery in order to accurately measure the front to back resistance of the diode. Meters with insufficient battery power may read infinite resistance (open) in each direction, mistakenly showing a good diode as being open.

The following resistance tests a diode and conclusively indicate a shorted diode. On physical inspection In most cases, defective diodes, whether shorted or open, will show some physical signs of the defect, such as a burned crack, a blistered spot, or it may even be split in two. Also, a shorted diode will usually give off a pungent burning odour.

Practical Exercise-Testing the HV Diode

Test Procedure for Testing HV Diode:

- Unplug the oven and open the screws and open the outer cover and locate the HV Diode.
- DISCHARGE ALL HIGH VOLTAGE CAPACITORS as per recommended procedure.
- Carefully remove the lead that goes to the capacitor (the ground connection may remain attached)
- Set the ohmmeter to read ohms at a scale of 10 K or higher.
- Measure the resistance in FORWARD BIAS across the terminals of the diode by touching the positive meter probe to the anode and the negative probe to the cathode (the cathode is the side that goes to ground, usually marked by an arrow, dot or stripe).
- A normal diode, depending on make and model, should read about 50,000 to 200,000 ohms.
- Reversing the leads (REVERSE BIAS) should produce a reading of infinity (open), unless there is a bleeder resistor across the diode, in which case the reading would show the [megohm] value of the resistor.
- If the meter shows continuity in both the directions, the diode is shorted. If infinity is read in both directions, the diode is open. In both the cases the diode should be replaced.

- In some models the diode is located inside of the high voltage capacitor. In this case, identify the diode terminal and perform the same test as above, measuring from the diode terminal to the capacitor's metal case.

High Voltage Transformer

HIGH-VOLTAGE TRANSFORMER as the name implies High Voltage to the unit is the POWERHOUSE of the microwave oven. With an input of 240 VAC applied to the primary winding, the high-voltage transformer (also referred to as power or plate transformer) steps up that primary voltage to a very high voltage. This high voltage is then boosted even higher by the voltage-doubling action of the capacitor and diode. The resulting voltage, about 3000 - 5000 volts DC (depending on the model), is available at the high voltage (output) tap (see illustration).

The transformer also has a second output winding which is stepped-down output that provides the filament voltage (typically 3 to 4 VAC) to the magnetron tube. It is dangerous to measure this voltage.

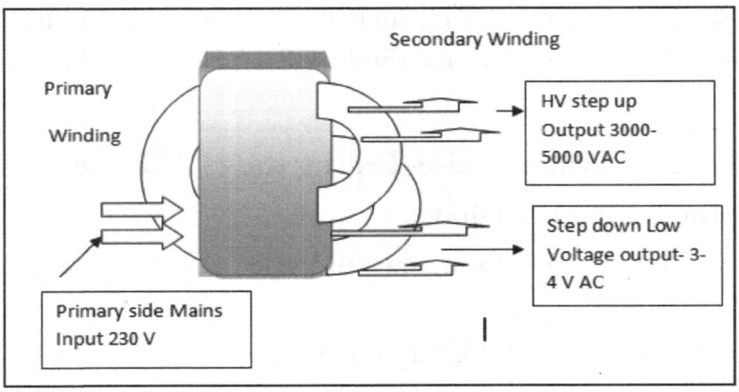

Testing the HV Transformer

Practical Exercise: Testing the HV Transformer

Caution: Before doing any test unplug the microwave oven and discharge all HV capacitors

1. UNPLUG the oven and open the oven.
2. DISCHARGE ALL HIGH-VOLTAGE CAPACITORS as per specified procedures.
3. Disconnect the high-voltage lead from the transformer's high-voltage terminal .
4. With an ohmmeter set to the lowest resistance scale, R X 1, measure the resistance from the high-voltage terminal to the transformer chassis which is grounded. For models having two or three high-voltage taps perform the measurement from each tap to chassis ground.
5. The meter should read about 55 to 70 ohms, depending on which high-voltage tap is being measured and the model being tested. A substantially higher or lower reading would indicate that the transformer is defective.

6. Carefully disconnect the leads from the primary (input) terminals.
7. Measure from one transformer primary terminal to the other for a normal reading of less than 1 ohm . A substantially higher reading would indicate that the primary winding is open.
8. Set the meter to its highest resistance scale and check from each transformer primary terminal to the transformer chassis (ground) for a normal reading of infinity (open circuit). Any measurement of resistance would indicate some degree of a short to ground
9. Set the meter back to its lowest resistance scale. Then carefully disconnect the filament leads and measure from one filament terminal to the other for a normal resistance of less than 1 ohm.
10. Set the meter to its highest scale and measure from each filament terminal to chassis ground for a normal reading of infinity (open circuit). Any measurement that is substantially different from the normal readings would indicate a defective transformer.

Magnetron

THE MAGNETRON is the heart of a microwave oven. It produces RF energy of 2.45 GHz and radiates the energy into the cooking cavity where it is absorbed by the molecules of the food. The magnetron uses permanent magnets and a half-wave voltage doubler circuit (the HV capacitor and HV diode) to oscillate and produce the 2450 Mhz cooking frequency, thus converting the 50 Hz supply voltage into microwave energy of 2.45 GHz.

Testing the Magnetron: The Magnetron condition can be found out from visual inspection and will require replacement. The following are the visual inspection symptoms that may suggest replacement of the Magnetron:

A small burnt spot on the magnetron insulator - indicates breakdown of insulation material. With subsequent uses it clearly shows visual evidence of the failure. It will produce Loud hum, no heat, arcing sound, electrical burning smell

The Magnetron will need replacement. The problem is depicted in the following picture:

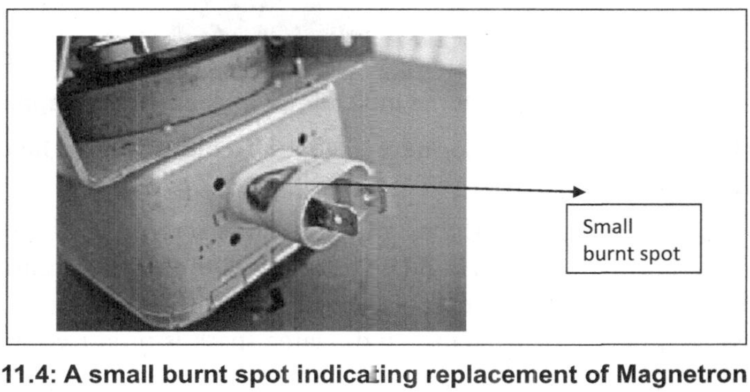

Figure 11.4: A small burnt spot indicating replacement of Magnetron

Cracked Magnet(s)- Will have symptoms of weak or no heat, magnetron gets extremely hot (overheats), intermittent arcing or "snapping" sound

Magnetron will need replacement

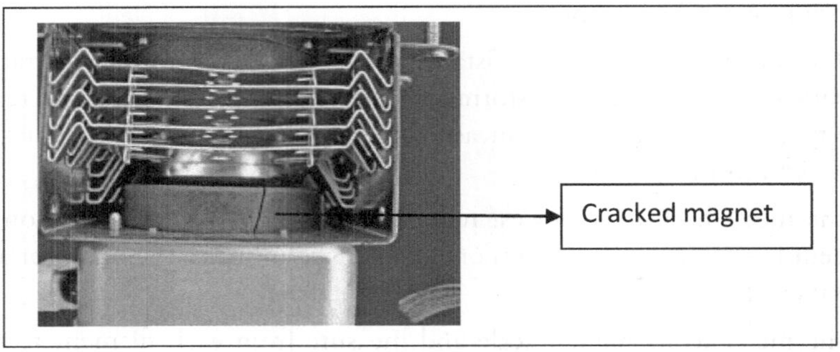

Figure 11.5: A cracked magnet in a magnetron indicating replacement of Magnetron

Burnt Dome (or Antenna) is caused by arcing due to back feeding microwave energy. When this occurs, check for a stalled or arcing stirrer blade or non-rotating antenna assembly. In units with multiple magnetrons, the cooking tray has to be raised in order inspect the condition of the lower antenna assembly. This also will show weak or no heat and arcing sound during the cook cycle.

In this case the Magnetron needs to be replaced and if damaged badly the antenna and stirrer assembly also need to be replaced.

Figure 11.6: Burnt Dome indicating replacement of Magnetron

Loose Magnetron Filament Connectors OR Discoloration of the connector(s) or plastic insulator(s):

If the connectors connecting the magnetron filament terminals become loose or are improperly crimped, it causes a build up of resistive heat. As this occurs the connection further deteriorates causing Small blackened pits in the magnetron terminal(s) and or Melted, and decayed appearance and a large spark is produced when discharging the capacitor. When this happens it will have effect of intermittent and/or low heat initially, then eventually no heat in the cavity.

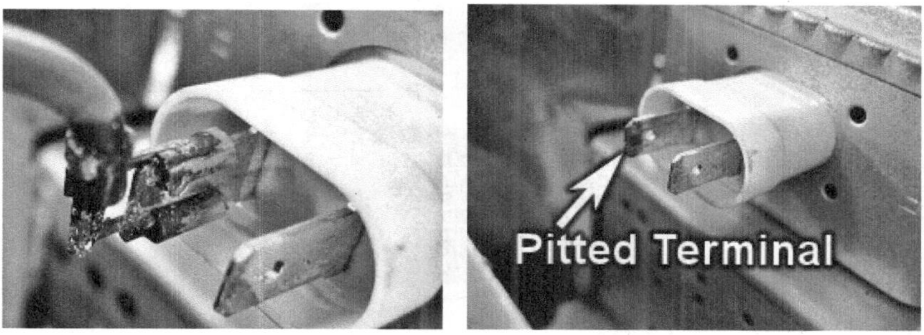

Figure 11.7: Burnt or pitted terminals or connector sheaths

In such cases the replacement of Magnetron will not be required. However repair defective terminals as follows:

Clean the burnt magnetron terminals and replace the slip-on connectors, making sure they fit tightly on the terminals; or Cut away burned wire and connector(s). Clean terminals to prepare and Solder the filament leads directly to the magnetron. Do not apply solderingheat for longer time if it is not necessary.

TESTING THE MAGNETRON

Practical Exercise: Testing and replacing a Magnetron

Caution: before doing any test unplug the microwave oven and discharge all HV capacitors

Unplug the oven.

DISCHARGE ALL HIGH VOLTAGE CAPACITORS as per specified procedures.

Make sure that all leads are removed from the magnetron terminals.

Testing for Filament Resistance (Low Resistance Test)

Keep the DMM to its lowest resistance scale.

Measure the resistance from one magnetron terminal to the other in either direction.

The magnetron filament resistance should be less than one ohm.

Testing for the short circuiting of Magnetron housing-(High Resistance Test)

Keep the DMM to highest resistance scale and check the resistance from the magnetron terminals to the metal magnetron housing.

The meter should show infinity oropen circuit regardless of meter polarity.

Even a slight reading would indicate a defective magnetron.

Replacing a Magnetron:

Figure 11.8: A typical Magnetron

The following should be kept in mind while replacing a Magnetron:

1. Do not touch the antenna dome area.
2. Be sure to transfer any add-on parts, such as an air duct or thermal fuse
3. Make sure that the wire mesh RF gasket is intact and in place
4. Examine the rim of the opening where the magnetron dome is to be inserted into the waveguide. Smooth out any irregularities, such as dents, pits, and burns. The rim surface should be bare metal, smooth to the touch. Use light-grade sandpaper - do not use steel wool.
5. If the terminal connections are found discolored, burnt, pitted connectors then repair or replace the slip-on connectors on the filament leads.
6. Perform an RF leakage check around the magnetron

High Voltage Capacitors

A combination of HV Diode and HV Capacitor is connected at the output of the High Voltage Transformer. This combination of Diode and Capacitor doubles the already High voltage present at the output of the HV Transformer. The capacitor can hold a fatal electrical charge long after the oven has been unplugged. Thus it is essential to Discharge this HV Capacitor before making any test or replacement.

TESTING HV CAPACITOR: BEFORE TESTING ANY CAPACITOR IT SHOULD BE DISCHARGED

Practical Exercise: How to discharge a HV Capacitor

Caution- before doing any test unplug the microwave oven and discharge all HV capacitor

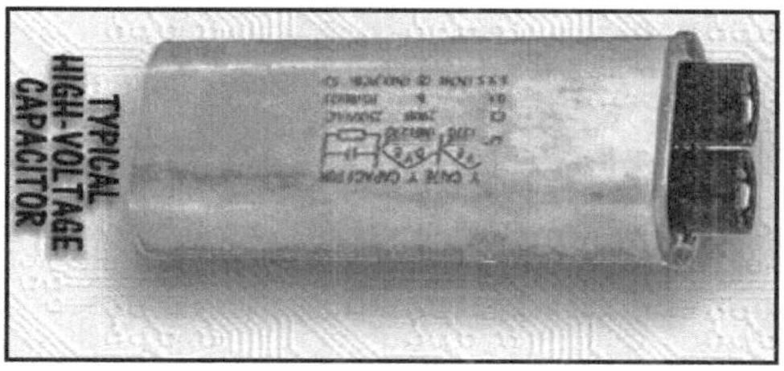

Figure 11.9: A Typical HV Capacitor

How to Discharge the Capacitor

1. ***Quick discharge method:***The capacitor is discharged by creating a short circuit (direct connection) between the two capacitor terminals and from each terminal to chassis ground (bare metal surface) with two screw drivers having effective insulated handle.

   a. Touch the blade of an insulated-handled screw driver to one terminal, then slide it toward the other terminal until it makes contact and hold it there for a few seconds. If the Oven was in use just before testing then the capacitor will be holding charges and will make a spark. And if there is Spark the capacitor is a good one.

   b. Repeat the procedure to create a short between each capacitor terminal and chassis ground.

2. ***Discharge with bleeder resistance:*** Attach one end of a 100K - 150K ohm, 25 watt resistor to the bare metal chassis with an alligator clip. Attach the other end of the resistor to the FLAT TIP of a well insulated screwdriver.

   Touch the TIP of the screwdriver to one of the capacitor terminals and hold it there for a few seconds. Then repeat the procedure for the other terminal.

   To ensure that the capacitor is fully discharged, follow the first method.

Figure 11.10: Quick Discharge Test

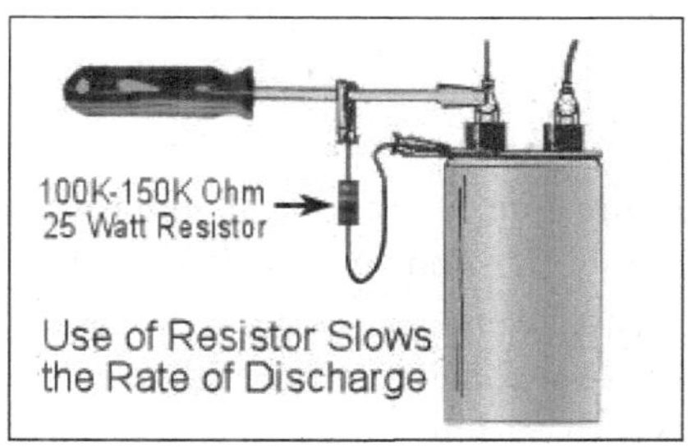

Figure 11.11: Discharge with bleeder resistor

Capacitor Test Procedure

1. Unplug the oven and open the same.
2. DISCHARGE ALL HIGH VOLTAGE CAPACITORS as per procedures.
3. Set the DMM to its highest resistance scale.
4. If the reading from one terminal to the other is infinity (or the value of the bleeder resistor) the capacitor is good.
5. Now reverse the leads. The meter should momentarily deflect toward the zero mark, then slowly drift back to infinity.
6. Reverse the leads once again. This should produce the same meter deflection.
7. Next measure from each terminal to the capacitor's metal case for a normal reading of infinity. (If there is an internal diode, the meter readings will reflect the diode's forward bias resistance. (See HV diode test procedure)
8. A visual inspection will also reveal certain defects, such as:
   - Evidence of arcing or burning at the insulators
   - The presence of an oily film or smell suggests a dielectric (non-conductive medium) leak
   - A bulging case indicates dielectric breakdown

Interlock Switch

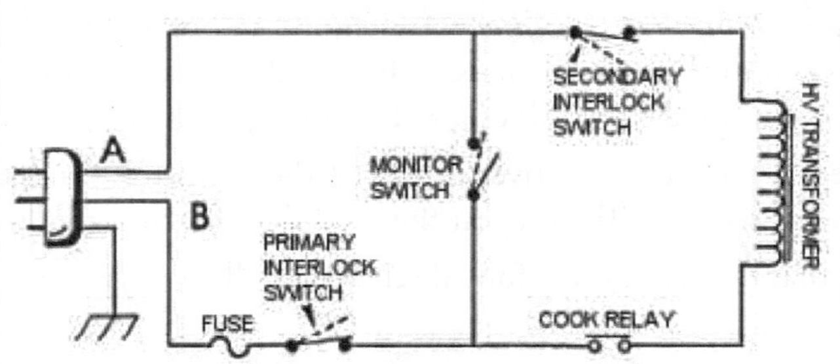

Simplified Interlock Monitor (Sensing) Circuit
(Shown with oven door closed)

To ensure safe operation, all microwave ovens are equipped with safety interlock switches . The door-interlock system is one of the most significant safeguards in a microwave oven. The purpose of the interlock system is to interrupt the production of microwave energy when the oven door is opened, and similarly, to prevent any microwave output until the door is firmly and safely closed.

A principal component of the interlock system is the interlock monitor switch . The monitor switch or the safety switch which disables the microwave oven if an interlock failure occurs.

The normal sequence of switch operation when the door is opened is as follows:

1. First the primary switch opens its contacts.
2. Second the secondary switch opens.
3. Finally, the interlock monitor switch closes its contacts.

Thus if any of the switches and/or relays included in the monitor loop (or circuit) fail to open their contacts properly when the door is opened, a short circuit is created when the monitor switch closes its contacts. The closed contacts of the monitor switch and the faultily-closed contacts of the defective switch combine to cause an immediate short circuit, which, in one way or another (depending on the model), blows the line fuse, or otherwise disables the oven. All this happens before the door can be opened far enough to allow any dangerous levels of microwave radiation to escape.

Testing: The interlock switch to be tested in this example has a COM. (common) terminal, a N.O. (normally open) terminal, and a N.C. (normally closed) terminal. Interlock monitor switches are usually constructed with only the COM and N.C. terminals. Other switches are made with just the COM and N.O. terminals. The following tests will cover most circumstances. Simply disregard the terminal-to-terminal tests that do not apply.

1. Unplug the oven and remove the outer cover.
2. DISCHARGE ALL HIGH VOLTAGE CAPACITOR as per specified procedure.
3. Physically examine the switch terminals and connectors for signs of overheating, such as discoloration, or brittleness.

   ❖ Many times the problem is merely a burned slip-on connector due to a poor crimp joint, or weakened connection. If the switch is in good working order, the repair can be made by cleaning the terminals, and replacing the burned connector. Or, simply cut off the burned connectors and solder the wires directly to the switch terminals.

   Carefully remove the harness leads from the switch terminals. Try them loose if necessary, but do not exert too much force, or the terminal may break right out of the switch.

   ❖ Many models use a connector with a locking clip in the center of the receptacle terminal. This extruding lever must be pressed down while gently pulling the connector off the terminal.

   Set the DMM to read ohms at a scale of R X 1 place one meter probe on the COM terminal and the other probe on the N.O. terminal. With the actuator (or lever, or button) not depressed, the meter should read infinity (an open circuit). Without moving the meter probes, press down on the switch actuator until a "click" is heard. At the point of the click, the meter should swing to a reading of zero ohms (or continuity). A healthy "click" usually means the switch is working normally.

TRIACS

The triac is another three-terminal ac switch that is triggered into conduction when a low-energy signal is applied to its gate terminal. Unlike the SCR, the triac conducts in either direction when turned on. The triac also differs from the SCR in that either a positive or negative gate signal triggers it into conduction. Thus the triac is a three terminal, four layer bidirectional semiconductor device that controls ac power whereas an SCR controls dc power or forward biased half cycles of ac in a load. Because of its bidirectional conduc¬tion property, the triac is widely used in the field of power electronics for control purposes. Triacs of 16 kW rating are readily available in the market.

"Triac" is an abbreviation for three terminal ac switch. 'Tri'-indicates that the device has three terminals and 'ac' indicates that the device controls alternating current or can conduct in either direction. Considering the heavy job that the triac does, it's not surprising that it is a common candidate for failure.

Symbol of Triac

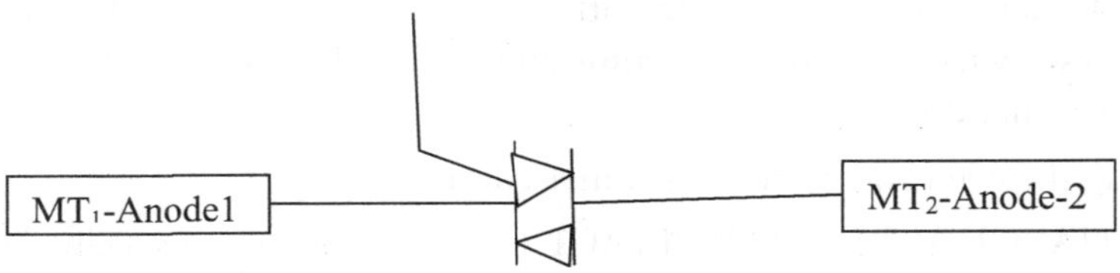

Testing the Triac:

http://www.microtechfactoryservice.com/safety.html- dischgTriacs with three terminals, such as most of those shown above, can be tested by making a series of resistance checks as follows.

1. ***Resistance measurement between Terminals, Gate and Ground***
   (i) Unplug the oven and open the same and locate the Triac.
   (ii) DISCHARGE THE HIGH VOLTAGE CAPACITOR as per specified procedures.
   (iii) Identify the terminals designated as G (gate), T1 and T2. (smallest terminal is the gate; medium sized is T1; largest is T2.)
   (iv) Carefully remove all harness leads. A soldered-in varistor or snubber may remain attached providing it's in good condition.
   (v) Set the DMM to a scale capable or reading LESS than 200 ohms.
   (vi) Take the reading from gate to T1 , note the reading.
   (vii) It should be between 10 to 200 Ohms.
   (viii) Note the reading again by reversing the leads. It should again be between 10 to 200 Ohms.

(ix) Set the DMM to the highest resistance scale. Measure the resistance between: T1 & T2; T2 & Gate; and between each terminal and Chasis ground.

(x) All the readings should be very high resistance or infinity or open circuit.

(xi) Any deviation would mean the Triac is bad and should be replaced.

2. ***Evaluate its gate-firing capability:***

(i) Unplug and open the oven.

(ii) DISCHARGE THE HIGH VOLTAGE CAPACITOR as per specific procedures.

(iii) Remove all harness leads. Set the meter to a scale capable of reading about 200 ohms.

(iv) Attach the negative meter lead to T1 and the positive lead to T2 .

(v) Now, using a screwdriver blade, create a momentary short between T2 and the gate . This brief contact should turn the triac "on," thus producing a meter reading of about 15 to 50 ohms.

(vi) Next, disconnect one of the meter leads, then re-connect it. The meter should return to a reading of infinity .

(vii) Finally, reverse the meter leads and repeat the tests. The results should be the same.

(viii) Any abnormal tests would suggest a defective triac.

(ix) Replace the defective triac.

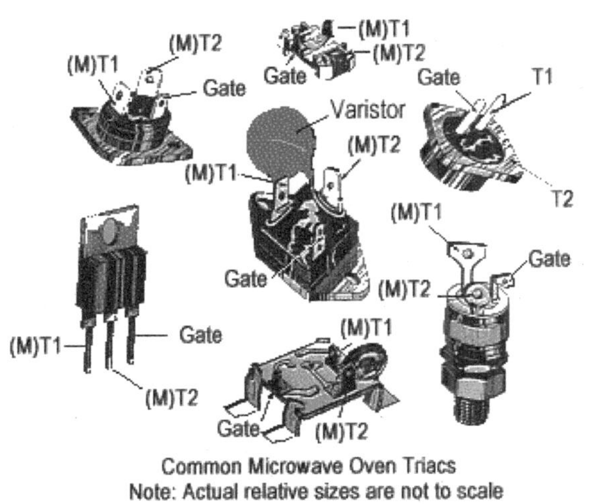

Common Microwave Oven Triacs
Note: Actual relative sizes are not to scale
(Artwork courtesy of Michael S. Wagner)

CHAPTER

12

Electric Iron

PRINCIPLE OF ELECTRIC IRON, PARTS OF STEAM IRON, THERMOSTAT HEAT CONTROLS

Electric Iron

Background

A clothes iron is a household appliance used to press the wrinkles out of and creases into clothes. When the iron is turned on, the consumer moves it over an item of clothing on an ironing board. The combination of heat and pressure removes wrinkles.

Most modern irons are made of metal and plastic, and have many features such as steam, temperature controls, and automatic shutoff. Steam provides an additional means for removing wrinkles from clothing.

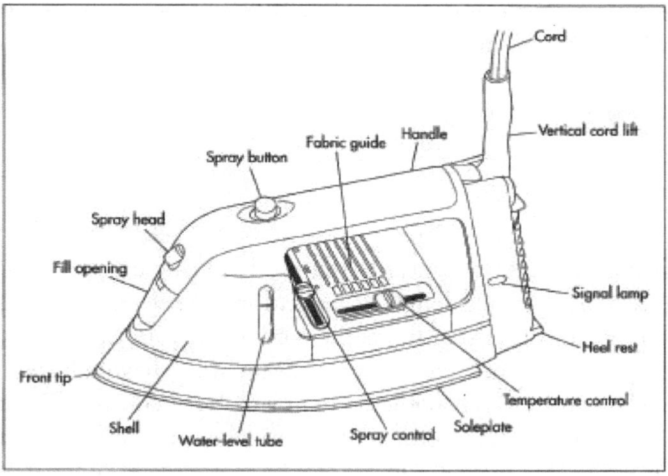

Figure 12.1: A typical clothes iron.

How Does a Steam Iron Work

Functions of Steam Iron

- A steam iron removes wrinkles from just about any fabric by application of heat, steam and weight. Most steam irons come with a setting for fabrics such as silk, polyester, wool and delicate materials, cotton and linen. Tough materials require the use of higher temperature settings, while the more delicate fabrics are best ironed at low temperatures. The heat, steam and weight of the iron stretch out the molecules in the material of clothing or cloth. Steam is usually reserved for tougher materials to stretch (e.g., cotton and linen).

  Heat and water resistance Sole Plate-

- The metal plate on the iron, commonly referred to as a sole plate, is usually made with aluminum. This aluminum plate has been manufactured with a water resistance treatment to the metal. The steam is created by releasing water from the water tank to the heated plate. The water is run through pores in the sole plate so that the water can be applied in a manageable amount. The steamed water is vaporized immediately after it is released from the pores in the sole plate.

Temperature Control of steam irons

- There is a thermostat embedded in the steam iron to ensure that a constant temperature will be produced when ironing any material. Users can set the appropriate temperature for ironing any material. Before ironing, simply select from a list of materials on the dial of the steam iron. The water tank has an indicator (or a see through plastic surface) for measuring the water level in the steam iron. When you're not sure about a setting for ironing a particular type of material, check the tag or flap on the product. Most clothes have a tag with instructions for washing, which includes the type of materials the clothes are made of.

Power

- Most steam irons come with an electrical power cord attached at the side or back end of the iron. The electrical power cord is made of a heat resistant insulation, which also prevents accidental electrocution. The electric power cord often comes with a built in spring that holds the cord away from the metal part of the steam iron for the operator's safety. Some steam irons have additional features. One grade of steam iron may come with an anti-burn control. The anti-burn control will shut off the power to the steam iron when it is left flat on an ironing board for too long. This prevents accidental fires. There are other steam irons that are completely cordless. These latter types of steam irons have a limit to the amount of power they can use or work with. Cordless steam irons are best for light loads of clothing that needs to be ironed. Still, some of the cordless steam irons come with a power saving feature. This power saving feature shuts off the power to the steam iron if it is left idle for a certain number of minutes. A sensor inside the steam iron's sole plate detects whether or not the sole plate is making contact with a material.

What Can Go Wrong with an Electric Iron?

There is very little inside a well-made iron that can go wrong. Most repairs are for faulty cords, damaged handles, and mineral deposits that hamper steam irons. Problems with an iron's electronic complements should be dealt with by an authorized service center. It typically is more cost effective to replace rather than repair an iron with internal problems. Check the owner's manual for your iron to learn the manufacturer's suggestions for cleaning, what water to use, and storage tips.

Soft-water systems add minerals that can harm an electric steam iron and your clothes so don't fill the iron with softened water. Instead, use filtered or distilled water.

How Can I Identify an Electric Iron Problem?

As with many small appliances, regular maintenance makes a dramatic difference in how trouble free your electric iron will be. Even so, things can happen.

- If the iron doesn't heat, make sure power is on to the outlet, check the electrical cord, and check the thermostat (see the Appliance Controls Fix-It Guide) and replace if necessary.
- If the iron heats but steams improperly, inspect the soleplate and clean the vents (see below) and flush sediment out of the steam chamber (see below).
- If the iron produces too much or too little heat, test the electrical cord. Also test and, if needed, adjust calibration of the thermostat (see the Appliance Controls Fix-It Guide).
- If the iron does not spray properly, inspect and clean the nozzle (see below).
- If the iron leaks or spits, clean the steam vents, nozzle, and tank.
- If the iron sticks to fabric, clean or repair the soleplate. If the iron stains fabric, clean the soleplate, clean the tank with a commercial cleaner, and use distilled or filtered water.

Caution!

Iron carefully around buttons, zippers, and other attachments or decorations that can scratch the soleplate.

What Do I Need for Electric Iron Repair?

Larger hardware stores may have replacement parts for popular brand electric irons. Also, you can get them from the manufacturer or an aftermarket supplier. Maintenance and repair tools you'll need to fix an electric iron include these:

- Screwdrivers
- Toothpicks or pipe cleaners
- Sewing needle
- Commercial electric iron cleaning solution or vinegar and water
- Commercial soleplate cleaner or baking soda and water
- Steel wool
- Emery cloth
- Metal cooking pot

What Are the Steps to Electric Iron Repair?

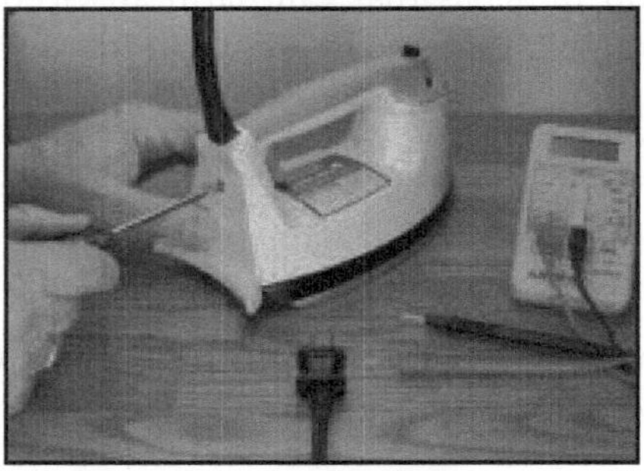

Access the internal parts of an electric iron by removing the rear cover panel. Some irons have unique fasteners to keep you from opening them. Check with your local hardware store for an appropriate screwdriver.

Clean an electric iron:

1. Unplug the iron and make sure it is cool before cleaning.
2. Use a toothpick or pipe cleaner to remove buildup in the steam vents, making sure the debris doesn't fall into the vents.
3. Use a fine sewing needle to carefully clean the spray nozzle of mineral deposits.
4. To flush sediment from a steam iron, pour 1/2 cup water and 1/2 cup vinegar into the water tank. Place the iron on a rack over a broiling pan and set the iron to steam until the tank runs dry. Repeat if necessary. Or follow the instructions for using a commercial iron cleaner.

Service an electric iron steam and spray mechanism:

1. Unplug the iron.
2. Use a fine sewing needle to unclog the steam valve assembly. Also, check the valve spring and replace it if it is broken or has lost tension.

3. If the spray pump is accessible, remove it and check for leaks by placing the spray tube in water and squirting the pump. Clean or replace as needed.

Clean an electric iron metal soleplate:

1. Unplug the iron.
2. Use a sponge and commercial soleplate cleaner or baking soda and water to remove dirt buildup on the soleplate. Rinse well with water and dry. Don't use harsh abrasives or immerse an electric iron in water.
3. Use very fine steel wool (0000) or an emery cloth to remove scratches and burns on the soleplate, then clean the soleplate.

Troubleshooting Tip

Unless the instructions with your iron say it's okay, don't let water stand in your steam iron between uses. Drain all water from the iron, wrap the cord loosely around the handle, and store the iron in an upright position.

CHAPTER

13 Rice Cooker

Principle of working of rice cooker.Various parts & functions of rice cooker, temperature control and timer unit.

THE RICE COOKER DIAGARM

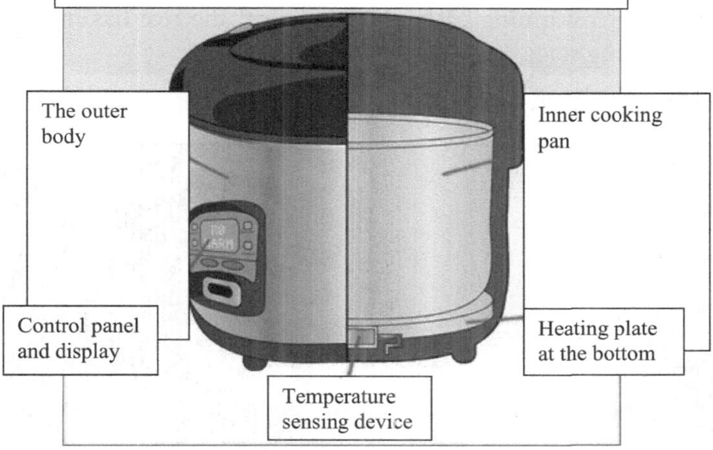

Rice Cooker Basics

Cooking rice happens in four phases:

1. Start of Heating
2. Boiling
3. Steaming
4. Resting

Rice cookers automatically guide rice through these four stages. The appliance consists primarily of a **outer body**, an **inner cooking pan**, an **electric heating plate at the bottom**, a **heat-sensing device** and some switches in various forms including buttons.

Water and rice sit inside the cooking pan while it's inserted into the rice cooker's shell. The pan's weight depresses the thermal-sensing device, and the heating plate quickly brings the water to a boil. The sensing device is a small, spring-loaded thermometer that gauges the temperature of the pan's contents. It's set into the bottom of the rice cooker's main body.

Simple rice cookers usually warm their contents by transferring heat from the heating plate to the cooking pan, and the type of metal used can improve that transfer. Some metals -- copper and aluminum for example -- are highly **conductive**. In other words, they transfer their heat easily. A wide range of materials can be used for the cooking pan, and each type may affect the overall time it takes to cook the food.

The process for cooking the rice is simple. Water boils at 100 degrees Celsius, and once it reaches a steady boil, it won't get hotter. As long as there is water in the pan, the temperature should be stable. Once the rice absorbs all the water in the pan, the temperature will start to rise. The rice cooker senses this change and will either switch off or switch to a warming cycle. At this point, the rice has finished cooking and entered the resting stage.

Repairing the Rice cooker

Figure 13.1: Rice cooker with base opened

Step 1: Unplug Power Cord from Wall

Before you start to work on the electric rice cooker you will want to make sure it is not plugged into any power source.

Step 2: Remove Cover

Take the covering dish of the electric rice cooker and place it aside. You will also need to remove the burner plate and inner bowl. This will allow you to get to the inner parts of the cooker.

Step 3: Remove Base of Cooker

On the bottom of most electric rice cookers there are some tabs, or screws, that hold the base together. Remove these screws, or fold back the tabs, so that you can remove the base.

Step 4: Test Heating Element

The cooker's heating element may not be working properly. Disconnect the heating wire from the terminal and test it with the multimeter. Set the meter to read Ohms and touch the wire to the terminals. If it reads "0" ohms the wire is fine.

Step 5: Test Resistor

If the heating element is fine, then the next space to test is the resistor. Use the multimeter again on the same setting and touch the probes to the resistor. If the reading comes out to 20 Ohms then the resistor is fine. If not, then you will need to replace it.

Step 6; Check and Clean Contacts

Another problem that is associated with older electric rice cooker appliances is that the switch contacts can become dirty or corrode. When this happens they will not allow a solid current to flow when the switch is pressed to star the cooker. With the base removed, check the contact areas. Check to see if they are burned. If so, you will need to replace these contacts. If not, then use some electrical contact cleaner and a clean rag.

Step 7: Move Contacts

If you notice that the contacts are not touching when you press the switch, you can fix this by using a small screwdriver and moving the contact closer to the switch.

Step 8: Replace Components

With the inside pieces either replaced, or cleaned, you can put the electric rice cooker back together. Make sure that the power cord is not pinched and that all wires are put back into their original position. Screw the base back on and replace the bowl and cover. Make a batch of rice to make sure it works correctly.

CHAPTER

14

Electric Kettle

Principle of working of electric kettle ,Various parts & functions of Electric kettle and temperature control unit.

PRINCIPAL OF WORKING OF ELECTRIC KETTLE

The electric kettle works on two key principle:

- The first is the immersed heating resistor, the piece of technology responsible for actually raising the temperature of the water in the kettle. Resistors, which take the form of the heating element in the bottom of the kettle, work by resisting the flow of electric current passed through them. This creates resistance and consequently heat. This heat is then passed into the water, which is then subsequently heated up.
- The second principle is automatic cut-off point, preventing the kettle from undesirably heating up the water. A bimetallic strip is introduced to the electric kettle , which, when heated by steam expanded, triggering a shut-off switch.
- Although some kettles have fancier and more complex heating and shut-off designs, it is through these two basic principles that the electric kettle works.

  When a current flows through the heating coil also known as element, water around the element is heated first by conduction and then the heat is spread all around by convection.

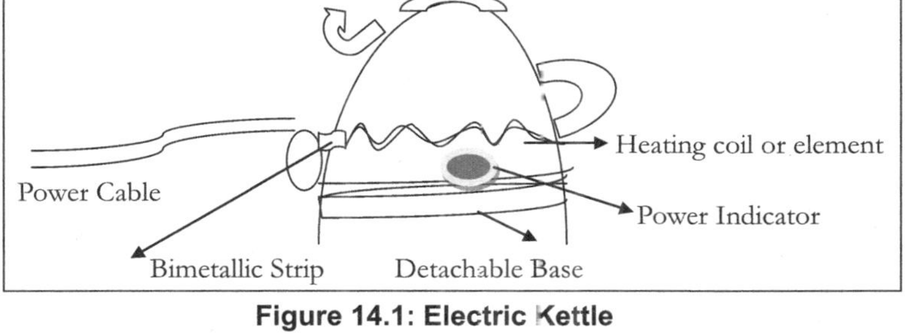

Figure 14.1: Electric Kettle

- ***Heating element:*** This works by resisting the flow of electrical current, which creates the heat the heats the water.
- ***Detachable base:*** A feature on all modern kettles, the base contains contacts that allow the flow of electricity to the element. (Not available in earlier kettles. The earlier kettles had integrated base through which the power supply was made as shown in the figure above)
- ***Power adapter:*** Connecting the heating element to the power supply allowing the flow of current through the element.
- ***Bimetallic strip:*** When the water heats up it causes the bimetallic strip to bend which triggers the switch that cuts off the power.

Troubleshooting and repair:

Problem	Repairs
The kettle will not switch on. The power indicator is not lighting.	Check the mains for 230 V with a meter. If this is ok then check the Fuse in the kettle. If the Fuse is not OK replace. If the Cable is detachable then check the voltage at the output of power cable. If the output is not there then replace the Cable.
The Kettle is not heating. However the power indicator is ON.	**The heating coil is burnt. Replace the same.**
The Heater will not cut off and will keep on heating.	The Heating cut off device- Bimetallic strip is damaged. Replace the same.

CHAPTER

15

Mixer Grinder

VARIOUS PARTS & FUNCTIONS OF MIXER/GRINDER, SPEED CONTROL CIRCUIT & AUTOMATIC OVER LOAD PROTECTOR

A mixer grinder is a very useful domestic appliance of the kitchen and is used to grind fruits, nuts, vegetables etc and to prepare delicious drinks like milk shakes. Dry grinding of spices, cereals, pulses, seeds, dry fruits etc and wet grinding of garlic, ginger, onion, vegetables etc are also done in mixer grinders. There are mixture grinders for mincing meat. Hotels and restaurants use high capacity mixer grinders. Different types of mixer grinders are also known as Food processor, juicer and grinder, food grinder, mixi, liquidizer etc.

Figure 15.1: Mixer Grinder

Numerous models of different capacities from various manufacturers are available. Now-a-days efficient models with attractive getup and added features are available and they make less noise and vibration compared to older models. Domestic mixer grinders work on 220 volt AC and commonly have power rating between 500 watt to 1.5 kW. Typical values are 500 watt, 750 watt, 1.1 kW etc. Maximum speed of 18000 to 20000 RPM is common. The main parts of a basic mixer grinder are:

1. Electric motor

2. Speed control switch
3. Overload protection and reset switch
4. Coupling
5. Jars and blades
6. Body
7. Accessories

Rotor or armature is also made of laminated steel

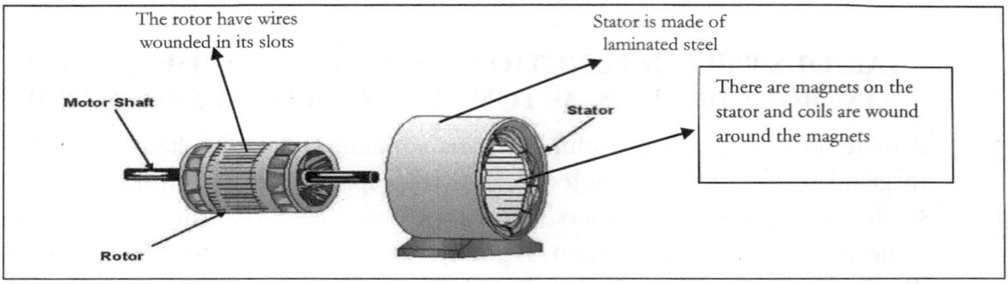

Figure 15.2: Motor

The motor used in mixer grinder is generally a universal motor. The motor has stator and rotor. The stator is made of laminated steel. There are poles on the stator. Usually two poles are there. Field coils are wound on each pole. The rotor, also called armature is also made of laminated steel, has slots and is tightly fitted with the shaft. Rotor windings are placed on rotor slots. Ends of rotor windings are permanently connected to commutator segments. Commutator segments are around the shaft on one side. There are two brushes that make contact with two commutator segments at a particular instant. As shaft rotates, the commutator segments that come in contact with the brushes also change. Field windings in the stator and rotor windings are connected in series through the rotor commutator. Universal motor can run on both AC and DC. They have high starting torque and speed.

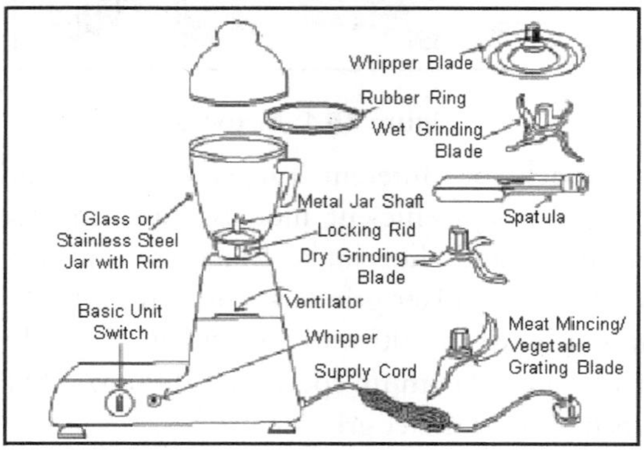

Figure 15.3: Parts of Mixer Grinder

The motor is fitted to the base unit with its shaft in vertical position. A top bowl contains the couplings that are tightly fitted on the top of the shaft where jars with blades are placed. The base of top bowl is sealed by rubber gaskets and is made leak proof Couplings are made of high quality food grade plastic or polycarbonate materials.

The speed control switch is generally a three speed switch. Low, medium or high speed can be selected by rotating the knob of the switch. There is overload protection for the motor. If the motor gets overloaded it automatically gets switched off. An overload reset switch is provided. Once the motor gets switched off because of overloading, the motor can again be started after pressing the overload reset switch.

Jars are made of stainless steel with transparent covers made of polycarbonate material. Coupling is at the bottom and blades are fitted inside. Base is sealed by rubber gaskets in such a way that even a drop of water can not leak through the base. The body is made of good quality plastic material, is insulated and shock proof. Accessories include jars of different capacity, blades of different types for dry grinding, wet grinding, vegetable mincing etc, rubber ring, spatula, power cord etc.

The jars are filled with food material to maximum two thirds of the jar volume and then placed on the top bowl of the mixer grinder. Proper coupling of the jar is ensured and jar cover is put on the jar. During operation jar should be kept securely held by holding the jar cover.

Common problems are defects in the motor or switch, worn out brushes, broken or loose blades, leak in top bowl or jars etc. The mixer grinder should not be continuously run for long periods. There should be a gap of minimum one minute between two successive switching on. The motor must be off and the shaft must be stationary at the instants of placing or removing the grinder or the jar.

Questions:

1. What type of motors is generally used in a Mixture grinder?
2. Mention two important features of the motor of a mixer grinder.
3. What is the function of overload reset switch in a mixer grinder?
4. State typical values of maximum speed of a mixer grinder motor.
5. How the field winding and rotor winding connected in the universal motor of a mixer grinder?

CHAPTER

16

Question Bank

BASIC ELECTRONICS- REPAIR & MAINTENANCE OF POWER SUPPLY, INVERTER & UPS

THEORY

Section-A

1. List any four causes for accidents.
   - Unawareness of danger
   - Disregard for safety
   - Negligence
   - Lack of understanding of proper safety procedures
   - Untidy condition of work place
   - Inadequate light and ventilation
   - Improper use of tools

2. What is the name given to the wire consisting of several strands of conductor?

Ans. Multistrand wire

3. What is the conventional direction of current flow in a DC circuit ?

   Conventional current is assumed to flow from +ve terminal to – ve terminal of the voltage source.

4. What is an earth electrode ?

Ans. A metal plate, pipe or a conductor electrically connected to the general mass of earth.

5. What is meant by an earth wire ?

Ans. The wires used for equipment earthing is called earth wire.

6. What are the devices used to protect a circuit from short circuit ?

Ans. Fuses, circuit breakers.

7. What is the rated output voltage of a nickel-cadmium cell?

Ans. 1.2 V

8. Name the semiconductor materials generally used to make transistors?

Ans. Silicon and germanium

9. How many characters are there in a hexadecimal system?

Ans. 16

10. What is the expansion of TTL family?

Ans. Transistor - transistor logic family

11. The shape of prohibition signs is ....

    A. triangular.

    B. circular.

    C. square.

    D. oblong.

Ans. B

12. The shape of warning signs is ....

    A. triangular.

    B. circular.

    C. square.

    D. oblong.

Ans. A

13. What is the action NOT to be taken in case a person is electruded?

    A. hold/catch the person with your bare hands.

    B. switch-OFF the electric power.

    C. remove the mains electric plug.

    D. insulate yourself and break the contact by pushing/pulling.

Ans. A

14. A multimeter cannot measure.....

    A. current

    B. potential difference

    C. capacitance

    D. resistance.

Ans. C

15. The polarity of test leads in a multimeter is not important in.....
    A. AC voltage measurement only
    B. DC voltage measurement only
    C. DC current only
    D. DC voltage and current measurement only.

Ans. A

16. Identify the AND gate...
    A. 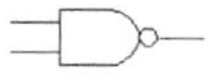
    B. 
    C. 
    D. 

Ans. D

17. When the ohmmeter probs are shorted, the meter should show..........
    A. zero ohms.
    B. infinite ohms.
    C 25 ohms.
    D 1K ohms.

Ans. A

18. Multimeter can measure
    A. V.I.A.R
    B. V,I
    C. A, R
    D. V.I.A

Ans. A

19. ______ meter displays measurement of dc or ac voltage in numerals.
    A. Analog voltmeter
    B. Digital voltmeter
    C. Digital ammeter
    D. Analog mutimeter

Ans. B

20. Wattage of the device is given by the product of ________ & __________
    A. I & T
    B. I & R

C. V & I

D. Q & V

Ans. C

21. Voltage is measured by using _____ meter

A. voltmeter

B. ammeter

C. watt meter

D. ohmmeter

Ans. A

22. Current is measured by using _______ meter

A. voltmeter

B. ammeter

C. watt meter

D. ohm meter

Ans. B

23. The value of the resistor for given colour code (Yellow Violet Yellow, Silver) is _______

A. 470 ohms

B. 474 ohms

C. 470 Kohms

D. 4.7 ohms

Ans. C

24. The combined resistance of the two 100 ohms resistors when connected in series _______

A. 200 ohms

B. 100 ohms

C. f300 ohms

D. zero

Ans. A

25. The polarity of an AC wave form reverses every ___ cycle

A. half

B. full

C. 1/8 th

D. 1/4th

Ans. A

26. The diode is a ____ terminal device

    A. 3

    B. 2

    C. 1

    D. 4

Ans. B

27. Full form of FET is ___

    A. Full effect transistor

    B. Fine effect transistor

    C. Field effect transistor

    D. Full emittor transistor

Ans. C

28. Full form of SCR is ___

    A. Solid crystal rectifier

    B. Silicon controlled rectifier

    C. Silver controlled rectifier

    D. 3Silver coated rectifier

Ans. B

29. How can you extend the backup time of an UPS system?

    A. increase the capacity of the battery.

    B. decrease the battery terminal voltage.

    C. change the type of battery.

    D. decrease the battery current rating.

Ans. A

30. An electrical circuit that has infinity resistance is called as .... circuit.

    A. an open

    B. a short

    C. a ground

    D. either B or C.

Ans. A

31. 7A Megger is exclusively designed for measuring ....

    A. very high resistance

B. very low resistance

C. ground faults in power lines

D. over loads on DC motors.

Ans. A

32. When test leads are open meter reading shows ....

A. 0

B. half scale

C. full scale

D. infinity

Ans. A

33. A rectifier is used to convert AC voltage into _________

A. A.C voltage

B. D.C voltage

C. AC current

D. DC current

Ans. B

34. A Megger measures resistance in ....

A. ohms

B. hundreds of ohms

C. thousands of ohms

D millions of ohms.

Ans. D

35. Match the statements with the labelled parts of the fig 4

1. Dial - multi-scale
2. Function selector
3. Pointer
4. Adjustment screw
5. Range selector switch
6. Makes it possible to adjust the pointer to zero position
7. Used to select different ranges of the meter
8. The position of D & E determines which one of it should be used to make correct reading
9. Set the meter to measure current, voltage (AC or DC) and resistance
10. Indicates measured value.

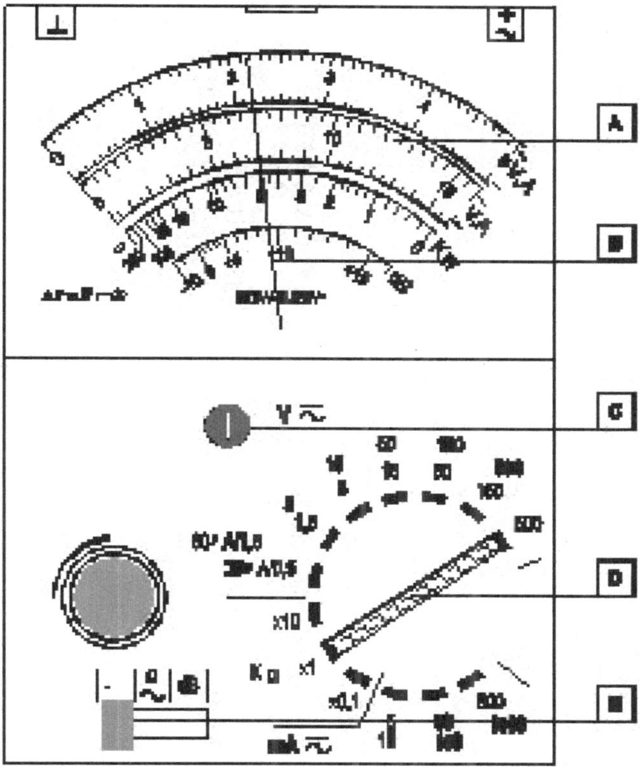

Ans. 1-A, 2-E, 3-B, 4-C, 5-D, 6-C, 7-D, 8-A, 9-E, 10-B

36. Identification of cells and batteries (Fig 1)

    1. Dry cell
    2. Lead acid battery
    3. Button cell
    4. Secondary cells
    5. Dry battery

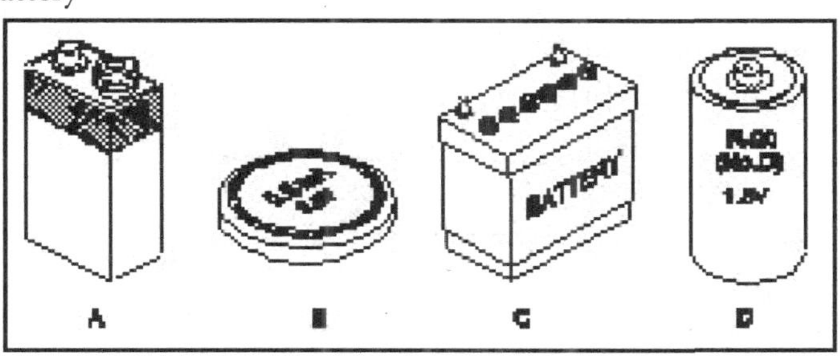

Ans. 1-D, 2-C, 3-B, 4-C, 5-A

37. Multimeter markings below Fig)

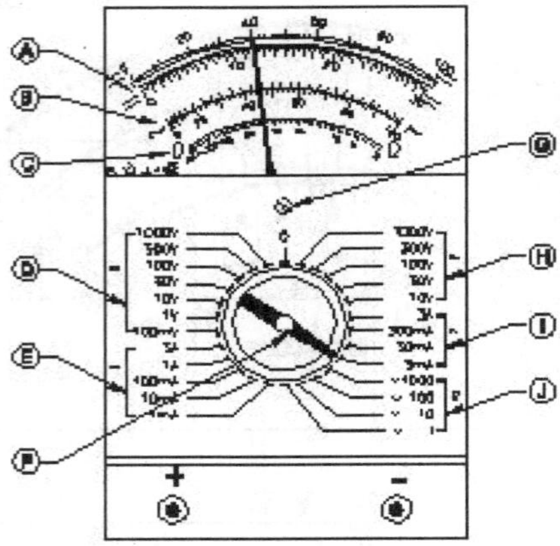

1. Mechanical zero setting
2. Scales for AC voltage and current
3. AC voltage range settings
4. DC voltage range settings
5. Scale for resistance
6. Scale for DC voltage and current
7. DC current ranges
8. Resistance range settings
9. Selection control knob/switch
10. AC current range settings

Ans. 1-G, 2-B, 3-H, 4-D, 5-C, 6-A, 7-E, 8-J, 9-F, 10-I

38. Transistor loadline parameters

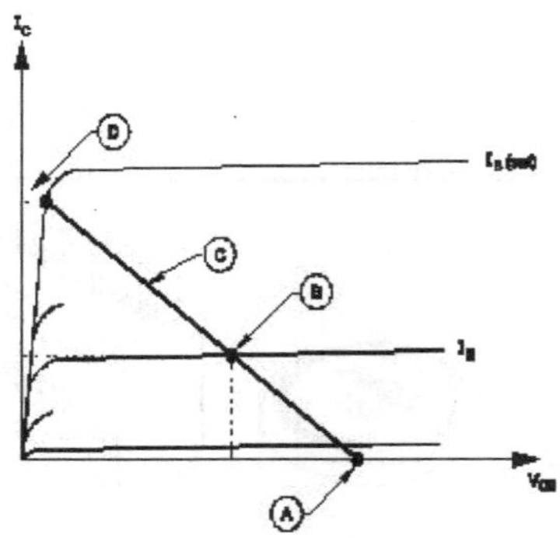

1. Load line
2. Cut off point

3. Quiescent point

Ans. 1-C, 2-A, 3-B

39. Transistor voltages

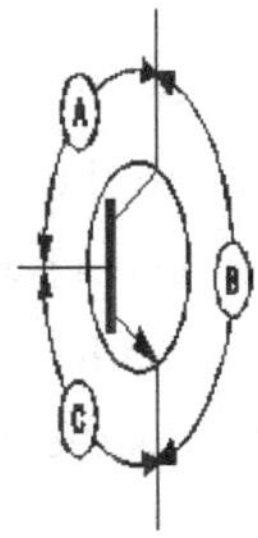

1 V_{CE}

2 V_{CB}

3 V_{BE}

Ans. 1-B, 2-A, 3-C

40 Match the description given below with the above mentioned TTL family IC

1 Package type

2 Device number

3 High speed CMOS TTL levels

4 Type of application

5. Identifies function

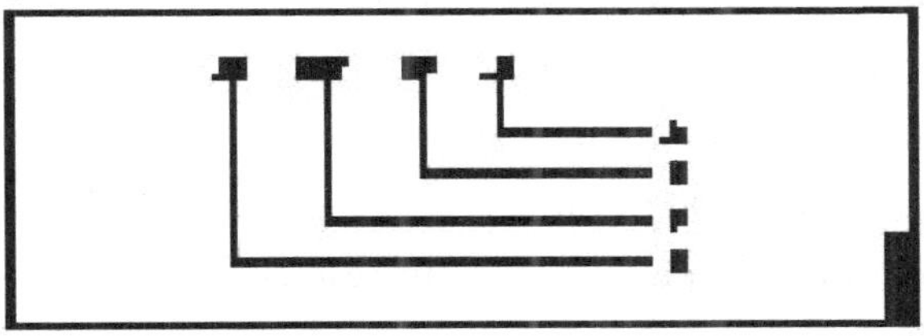

Ans. 1-A, 2-B, 3-C, 4-D, 5-B

Section - B

1. List the four basic categories of sign boards.

Ans.

(i) Prohibitory signs

(ii) Mandatory signs

(iii) Warning signs

(iv) Information signs

2. Name any two instruments commonly used for measuring the outer dia of wires?

Ans. (i) standard wire gauge

(ii) outside micrometer (micrometer calipers)

3. What is the direction of electron flow in a DC circuit ?

Ans. Electron flow is from – ve polarity to the +ve polarity of a voltage source.

4. What are the effects of open circuit in a series circuit ?

Ans. (a) No current flow through the circuit.

(b) Full supply voltage appear across the open.

(c) Power consumption is zero

5. How an analog signal of 3 volts represented in digital value?

Ans. 0011

6. What is the logical operation used for a circuit to get inverse output of the input signal?

Ans. NOT

7. How many PN junctions are there in a SCR?

Ans. Two

8. What is the expansion of MOSFET?

Ans. Metal oxide semiconductor field effect transistor

9. First aid/artificial respiration recovers a victim temporarily from his sickness. What is the immediate step you have to do?

A. allow him to do his work/duty.

B. call a doctor for treatment.

C. give heavy/solid food to eat.

D. allow him to do some body warming exercise.

Ans. B

10. The severity of an electric shock depends on many factors. Which factor NOT influences the severity of shock?

A. The level of current.

B. The duration of current passing.

C. The person who receives the shock.

D. Voltage level of electricity.

Ans. C

11. Electron volt is the unit of
    A. voltage
    B. current
    C. energy
    D. ohms

Ans. C

12. Different cores used in transformers are iron & _____
    A. air
    B. mica
    C. ferrite
    D. paper

Ans. C

13. Rating of a battery is usually expressed in ____
    A. ampere-watt
    B. ampere-hour
    C. ampere-ohm
    D. volts

Ans. B

14. A zener is operated in the _____ region
    A. forward
    B. reverse
    C. middle
    D. high

Ans. B

15. The no. diodes used in half wave rectifier are _______
    A. 2
    B. 4
    C. 1
    D. 3

Ans. C

16. The no. diodes used in full wave bridge rectifier are _______
    A. 2
    B. 4

C. 1

D. 3

Ans. B

17. A bipolar junction transistor consists of base, collector & ______

A. emitter

B. control grid

C. gate

D. anode

Ans. A

18. SCR is _____ terminal device

A. 4

B. 3

C. 2

D. 1

Ans. B

19. Which output waveform from UPS offer a better performance?

A. quasi-square wave.

B. square wave.

C. sine wave.

D. saw tooth wave.

Ans. C

20. What will be the symptom if there is a failure in PWM or feedback section of an ON line UPS?

A. output voltage will jump above 300 V.

B. output voltage will fall to zero.

C. output voltage will not change.

D. output frequency will fluctuate.

Ans. A

21. In a line interactive UPS, the battery charging voltage is provided by…

A. AC mains

B. inverter

C. transfer switch

D. current transformer.

Ans. B

22. When using a Megger you should avoid touching ....
    - A. the positive lead
    - B. the negative lead
    - C. both leads at the same time
    - D. either lead.

Ans. D

23. A victim having wet skin who is in contact with a live wire of 240 V could not let the wire to go from his hand. In this case, the minimum current through the body will be ....
    - A. 5 mA
    - B. 10 mA
    - C. 20 mA
    - D. 50 mA.

Ans. D

24 The purpose of the capacitor used at the input side of a three terminal regulators is to ....
   - A. filter the ripple.
   - B. prevent the oscillations.
   - C. improve the ripple rejection ratio.
   - D. block dc current.

Ans. B

25 Safety sign boards - meanings

1. Naked film prohibited	A 
2. Risk of electric shock	B 
3. Toxic hazard	C 
4. Wear tool protection	D 
5. General warning risk of danger	E 

Ans. 1-B, 2-D, 3-A, 4-E, 5-C

26. Materials - Classification

1. Copper	A	Insulating material
2. Silver	B	Semi-conductor
3. Porcelain	C	Good conductor

4 Germanium D Best conductor

5 Silicon

Ans. 1-C, 2-D, 3-A, 4-B, 5-B

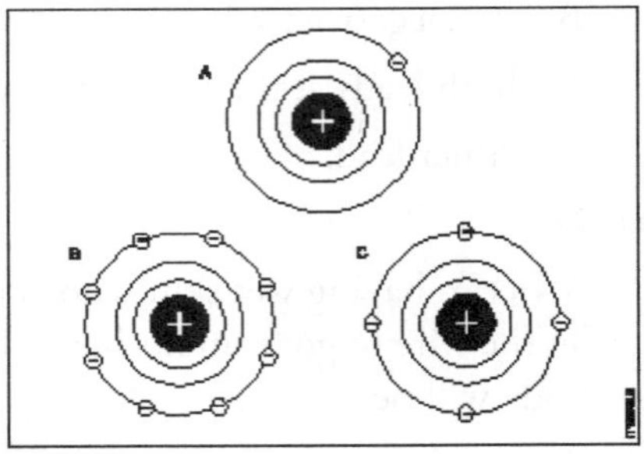

27. Material property - Valence electrons
    1. Semiconductor
    2. Conductor
    3. Insulator

Ans. 1-C, 2-A, 3-B

28. Transistor code numbers Types

1.	Silicon,RF transistor	A	AC127
2.	Germanium, AF transistor	B	BF195D
3.	Silicon, switching high power transistor	C	AD128
4.	Germanium power transistor	D	BU208

Ans. 1-B, 2-A, 3-D, 4-C

29. Match the following:

1.	$(44)_{10}$	A	$(110111)_2$
2.	$(35)_{10}$	B	$(101100)_2$
3.	$(55)_{10}$	C	$(100011)_2$
4.	$(2C)_{16}$		

Ans. 1-B, 2-C, 3-A, 4-B

30. Match the correct output for the positions marked in the Fig.
    1. A B
    2. B A
    3. A
    4. B

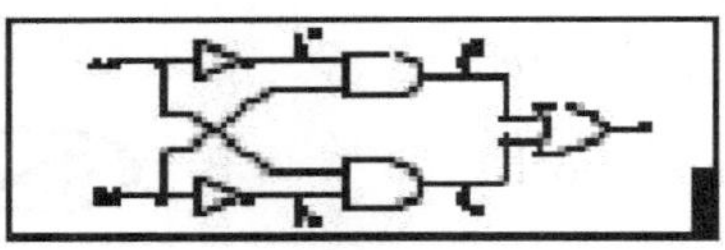

Ans. 1-D, 2-C, 3-B, 4-A

Section - C

1. Which method should be used while giving artificial respiration to the patients who have injuries in the chest wall and abdomen?

Ans. Holgen-Nelson method, Schater's method

2. Identify the method of artificial respiration illustrated in the Fig

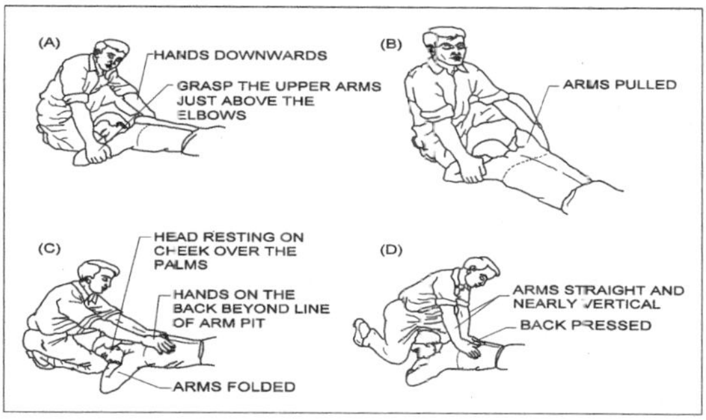

Ans. Nelson's method

3. Refer to Fig 1, if the fuse does blow and opens the circuit what will be the voltage drop across R ?

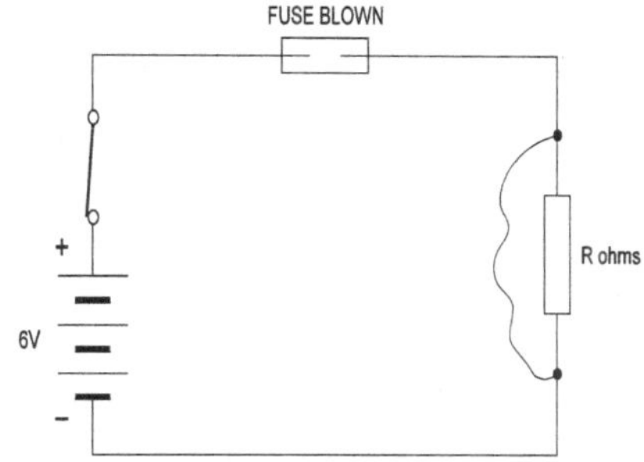

Ans. Voltage drop across the resistor will be zero.

4. What is the reading of a voltmeter connected across the resistors R1, in the circuit shown in fig 2 ?

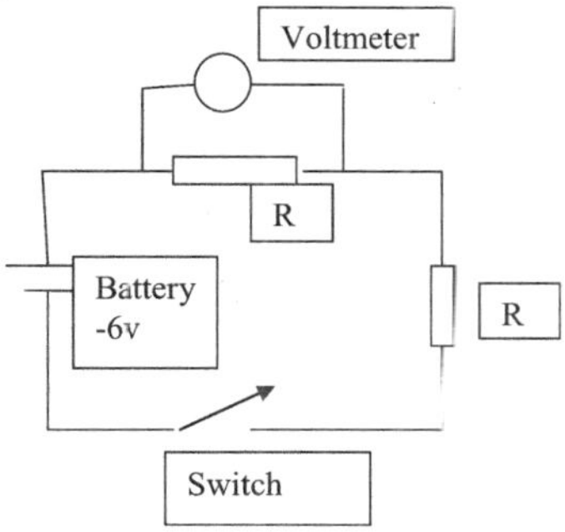

Ans. Voltmeter indicates 6 volts.

5. What are the different classification of counters?

Ans. Synchronous counters and asynchronous counters

6. In kinetic method of lifting an object, the feet should be placed about ....

A. 6 inches apart.

B. 10 inches apart.

C. 18 inches apart.

D. 36 inches apart.

Ans. C

7. The requirements of combustion are ....

A. fuel and oxygen.

B. fuel and heat.

C. heat and oxygen.

D. fuel, heat and oxygen.

Ans. D

8. For proper functioning of a transistor its emitter base junction is always forward bias and collector base junction is...

A. Forward biased

B. Reverse biased

C. Not biased

D. Shooted

Ans. B

9. The majority carriers in NPN transistors are ....

A. holes

B. ions

C. electrons

D. charge

Ans. C

10. The majority carriers in PNP transistors are ....

A. holes

B. ions

C. electrons

D. charge

Ans. A

11. What is the minimum change over time allowed in a computer to prevent its reboot in case of transfer of load takes place in an OFF line UPS?

A. less than 3 ms.

B. less than 5 ms.

C. less than 7 ms.

D. less than 10 ms.

Ans. B

12. Which of the following is an essential component in an OFF line UPS?

A. Relay.

B. Timer.

C. Counter.

D. Thermostat.

Ans. A

13. Which device is protecting the UPS by blowing the fuse in the case of high I/P voltage?

A. HRC fuse.

B. Metal oxide varistor.

C. Thermistor.

D. MOSFETS.

Ans. B

14. What is the wave shape in a pulse width modulated output voltage of an OFF line UPS?

A. square wave.

B. sine wave.

C. saw tooth wave.

D. quasi square wave.

Ans. A

15. The IC 3525 used in OFF line UPS is an...

A. amplifier.

B. regulator.

C. oscillator.

D. detector.

Ans. C

16. What is the formula to find output load of an UPS?

A. volt x current.

B. current x load resistance.

C. load resistance x voltage.

D. volt ampere x power factor.

Ans. D

17. What is the maximum output load of a 2.5 KVA UPS having a power factor of 0.7?
    A. 500 W.
    B. 750 W.
    C. 1250 W.
    D. 1750 W.

Ans. D

18. What is the percentage value below the rated value considered to be the low battery in a UPS?
    A. 5%.
    B. 10%.
    C. 15%.
    D. 20%.

Ans. B

19. In an OFF line UPS, if the mains input exceeds a preset value, the UPS is forced to changeover to battery mode. This is known as...
    A. mains low cut.
    B. load set point.
    C. mains high cut.
    D. load cut-off point.

Ans. C

20. Which device can be used to sense the minimum value of mains input voltage in a typical UPS?
    A. thermister.
    B. optocoupler.
    C. varister.
    D. varicap.

Ans. B

21. How can you increase the efficiency of an ON line UPS?
    A. increase the rated battery voltage.
    B. increase the rated battery current.
    C. use boost type power factor connection.
    D. use always constant load.

Ans. C

22. ON line UPS can maintain a constant charging current. The charging current depends on

    A. load current of the circuit

    B. terminal voltage of the battery

    C. UPS capacity/rating

    D. ampere hour of battery.

Ans. D

23. How do you maintain constant charging current in an ON line UPS?

    A. controlling input voltage to SCR.

    B. controlling terminal voltage of UPS.

    C. controlling firing angle of SCR.

    D. controlling load current.

Ans. C

24. UPS system can be used in combination with AC generators and the selection should be in accordance with the load. Which current we have to consider?

    A. average current.

    B. peak current.

    C. quarter value of load current.

    D. half of the load current.

Ans. B

25. If the output terminals of the regulator is short circuited, the current flow through the current sensing resistor will be ....

    A. zero.

    B. low.

    C. high.

    D. normal.

Ans. C

26. The digital circuit that routes its single input signal to one of 2^n output lines depending on the values of n control lines is...

    A. multiplexer

    B. demultiplexer

    C. encoder

    D. data selector.

Ans. B

27. Types of feedback

1. +ve feedback	A	De-generative feedback
2. –ve feedback	B	Non-inverting feedback
3. Current series feedback	C	Re-generative feedback
4. Voltage series feedback	D	Inverting feedback

Ans. 1-C, 2-A, 3-B, 4-D

28. Types of bias

1. Base bias
2. Emitter bias
3. Voltage divider bias

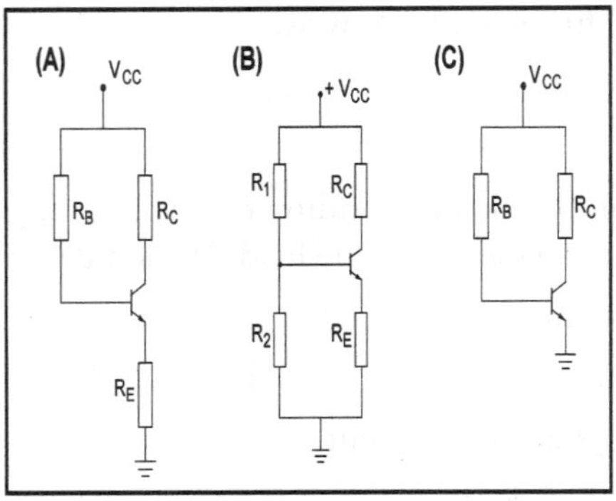

Ans. 1-C, 2-A, 3-B

29. Transistor leads -functions

1. Controls flow of current carriers	A	Emitter
2. Collects current carriers	B	Base
3. Sometimes connected to metal casing to dissipate heat	C	Collector
4. Emits current carriers		

Ans. 1-B, 2-C, 3-C, 4-A

30. Transistor currents Magnitudes

1. I_C	A	nanoamperes
2. I_{CO}	B	microamperes
3. I_B	C	milliamperes
4. I_E		

Ans. 1-B, 2-C, 3-C, 4-C

BASIC ELECTRONICS- REPAIR & MAINTENANCE OF POWER SUPPLY, INVERTER & UPS

PRACTICAL

Section – A

Time: 2 hours

1. a) Prepare a sign board
   b) Identify different types of hand tools (minimum 10 numbers)
2. a) Measure the DC voltage and DC current in a circuit using multimeter
   b) Find the value of the given resistors (minimum 10 numbers) by colour code and by using multimeter.
3. a) Soldering and desoldering the given components in a given PCB
      i) Resistors (5 nois)
      ii) Diodes (5 nos)
      iii) Transistors (5 nos)
   b) Test the regulated and unregulated voltages of a given circuit.
4. Construct and test a Full wave and a Bridge rectifier circuit.
5. a) Construct and test a Half-wave rectifier.
   b) Identify and test the given components.
      (Assorted electronic components minimum 10)

Section – B

Time: 4 hours

1. Assemble and test an audio power amplifier.
2. a) Assemble and test a CE amplifier.
   b) Assemble a RC-oscillator and test it.

3. Assemble and test an Inverter circuit.
4. Find and rectify the faults in the given UPS.
5. Find and rectify the faults in a given SMPS

MODULAR EMPLOYABLE SKILLS

QB FOR ELECTRONICS SECTOR

REPAIR & MAINTENANCE OF WASHING MACHINE & MICROWAVE OVEN

THEORY

Section – A

Choose the correct answer

1. What the below picture depicts?

a. bowl b. microwave oven c. fax machine d. washing machine

Ans. d

2. The capacity of washing machine is given by

a. kilogram b. Newton c. joules d. tones

Ans. a

3. The speed at which the spin tub rotate is given by

a. m/sec b. rps c. rpm d. km/sec

Ans. c

4. At what speed the spin basket rotates?

a. 200 – 300 rpm b. 100 – 300 rpm

c. 700 – 900 rpm d. 800 – 900 rpm

Ans. d

5. Which creates the washing action?

a. turbulence b. energy c. electricity d. water supply

Ans. a

6. Which part of washing machine removes the dirt from clothes?

a. wash lid b. spin lid c. pulsator/agitator d. wash time

Ans. c

7. To which the washed clothes are transferred for drying purpose

a. twin tub b. spin tub c. agitator d. single tub

Ans. b

8. How much amount of water is removed from the clothes by spin tub?

   a. 50% – 80% b. 10% – 50% c. 40% – 50% d. 30% – 80%

Ans. a

9. The remaining amount of water can be removed by

   a. left within the washing machine itself
   b. hanging out in the sun light/open air
   c. placing it in the tub
   d. again washing it

Ans. b

10. What is the maximum amount of program to be stored in the fuzzy logic machine?

    a. around 1000 b. around 500 c. around 2500 d. around 2000

Ans. d

11. In what position the pulsator to be placed in the above type washing machine?

    a. middle of tub b. front of tub c. back of tub d. bottom of tub

Ans. d

12. Agitator in the washing machine looks like

    a. leaf b. bowl c. stem d. fan

Ans. c

13. In what position the agitator to be placed in the wash tub?

    a. middle of tub b. front of tub c. back of tub d. bottom of tub

Ans. a

14. With what material the wash and spin leads are made of?

    a. iron b. steel c. plastic/fiber d. copper

Ans. c

15. Which timer is used for fully automatic washing machines?

    a. mechanical timer b. spring timer

    c. electric timer d. electronic timer

Ans. d

16. Which is used to control the water level in the automatic washing machine?

    a. tub b. solenoid valve c. tap d. motor

Ans. b

17. Which is used to produce the air bubbles to wash clothes?

    a. water level controller b. pulsator

    c. wash tub d. air bubble generator

Ans. d

18. Which type of motor is used in washing machine?

    a. DC motor b. synchronous motor

    c. capacitor run motor d. induction motor

Ans. c

19. What is the speed at which the motor runs for washing?

    a. 200 rpm b. 100 rpm c. 150 rpm d. 145 rpm

Ans. d

20. In which mode the power consumption would be less?

    a. heavy mode b. economic mode

    c. soak mode d. fuzzy mode

Ans. b

21. In which form the fully automatic washing machine is available?

    a. 4 kg b. 10 kgc. 5 kg d. 3 kg

Ans. c

22. Which mode takes long time to wash clothes?

    a. light mode b. economic mode

    c. soak mode d. fuzzy mode

Ans. c

23. How many springs are needed to connect the tub to the main body?

    a. 1 b. 3 c. 8 d.4

Ans. 4

24. Which is to be varied on the sensor tube to change the volume of water in the tub?

    a. temperature b. force c. energy d. pressure

Ans. d

25. How many layers are there in the wash tub?

    a. 1 b. 2 c.3 d. 4

Ans. b

26. Which is the good reflector of microwaves?

    a. cavity b. glass c. paper d. wood

Ans. a

27. How much power does microwave oven consumes?

    a. 100 – 200 W b. 300 – 500 W

    c. 500 – 1000 W d. 1000 – 2000 W

Ans. c

28. What is the frequency used to heat the food using microwave oven?

    a. 2.45 Hz b. 2.45 KHz c. 2.45 MHz d. 2.45 GHz

Ans. d

29. What is the power level cycle used in microwave oven?

    a. 10 – 30 sec b. 1 – 10 sec c. 1 – 5 sec d. 1 – 15 sec

Ans. a

30. What is the warm up period after microwave power is switched on?

    a. 0.5 sec b. 1 sec c. 1 – 3 sec d. 1 – 2 sec

Ans. c

31. What is the capacity of basic model microwave oven?

    a. 10 – 40 Lt b. 5 – 10 Lt c. 50 – 80 Lt d. 20 – 60 Lt

Ans. a

32. What is the speed of the turn table motor?

    a. 5 – 20 rpm b. 10 – 20 rpm c. 15 – 20 rpm d. 5 – 10 rpm

Ans. d

33. What is the working voltage of turn table motor?

    a. 440 V AC b. 230 V AC c. 220 V AC d. 400 V AC

Ans. b

34. What is the power consumed by turn table motor?

a. . 5 – 20 W b. 10 – 20 W c. 15 – 20 W d. 5 – 10 W

Ans. d

35. Which is the most dangerous of consumer appliances?

a. iron box b. microwave oven

c. washing machine d. mixer

Ans. b

36. Which is the energy storage device used in microwave oven?

a. capacitor b. inductor c. resistor d. transistor

Ans. a

37. What is the resistor value used to dissipate the high voltage in microwave oven?

a. 20 W b. 10 W c. 25 W d. 30 w

Ans. c

38. What does magnetron consists of in microwave oven?

a. heated filament anode b. transformer

c. heated filament cathode d. transducer

Ans. c

39. What is the working voltage of HV capacitor in microwave oven?

a. 2 M V AC b. 2 K V AC c. 1 M VAC d. 1 K VAC

Ans. b

40. Which rotates along the food during the cook cycle?

a. grill b. thermal fuse c. thermostat d. turn table

Ans. d

41. Which motor controls a door that is normally closed during cooking time?

a. servo motor b. synchronous motor

c. damper motor d. induction motor

Ans. c

42. What is the main use of combinational model microwave oven?

a. greening the food b. browning the food

c. reddening the food d. boiling water

Ans. b

43. Which is used as a part of HV capacitor to discharge the capacitor?

a. bleeder resistor b. inductor c. transformer d. transducer

Ans. a

44. Which is the electrical conventional heat coil in microwave oven?

    a. grill   b. thermal fuse   c. thermostat   d. turn table

Ans. a

45. What is the rating of door switches?

    a. 440 V/ 10 A   b. 230 V/10 A   c. 230 V/15 A   d. 400 V/ 15 A

Ans. c

46. Which is used to control the switching of HV source?

    a. relays   b. thermal fuse   c. thermostat   d. turn table

Ans. a

47. Which supplies the voltage to magnetron?

    a. high voltage transformer   b. HV diode

    c. thermostat   d. HV capacitor

Ans. a

48. By what way the heat is produced in basic model microwave oven?

    a. electric power   b. microwave power   c. battery   d. transformer

Ans. b

49. Which model is available in micro controlled models?

    a. basic model microwave oven

    b. combinational model microwave oven

    c. convention model microwave oven

    d. electric model microwave oven

Ans. c

50. By which the microwave generating circuit is activated by the controller?

    a. triac   b. diode   c. fuse   d. capacitor

Ans. a

Section – B

Write short answers (One or two words)

51. Which part of washing machine is used to remove the water from cloth?

Ans. Spin

52. In front loading machine, in what axis the drum rotates?

Ans. Horizontal axis

53. Which material is used to make tub?

Ans. Molded polypropylene

54. Which part of washing machine is used to rotate the water?

Ans. Pulsator

55. By what means the wash motor is connected to the pulsator pulley?

Ans. "V" belt

56. In which mode heavy clothes like jeans or blanket are washed?

Ans. Heavy mode

57. In which mode normal day – to day clothes like cotton material is washed?

Ans. Economy mode

58. In which mode soaking is done?

Ans. Soak mode

59. Which is called as the brain of washing machine?

Ans. PCB

60. Which part is connected between the water outlet and room drain?

Ans. Drain hose

61. Which is used to control the time of the spin/rinse process?

Ans. Spin timer

62. Which is used to smash clothes?

Ans. Agitator

63. What is the rating of HV capacitor in microwave oven?

Ans. 0.65 – 1.2 μf

64. What does micro stands for?

Ans. Microwave cooking

65. What does the controller includes in microwave oven?

Ans. Microcomputer

66. Which is used for the safety in microwave oven?

Ans. Door switches

67. Which is used to measure the temperature?

Ans. Thermostat

68. Where the user operational switches are present in microwave oven?

Ans. Key pad

69. Which is used to rectify the output of HVT and supply to magnetron?

Ans. HV diode

70. What is the voltage present in the secondary of HVT?

Ans. 2 K V RMS

71. What is the low voltage winding for magnetron filament?

Ans. 3.3 V

72. What is the RC time constant in microwave oven?

Ans. 0.1 sec

73. Which is used to cool the magnetron?

Ans. Cooling fan motor

74. What is the amount of HV may exceed?

Ans. 5000 V

75. What is the power rating of the HV transformer may exceed?

Ans. 1500 W

Section – C

Write short answers (One or two sentence)

76. Why microwave oven is dangerous?

Ans. Very high voltages (upto 5000 V) at potentially very high currents are present

77. What is the use of Thermal fuse?

Ans. It is blown when the temperature of magnetron goes beyond the limit.

78. What does micro stands for?

Ans. Microwave cooking

79. What is the use of magnetron?

Ans. Used to produce microwave signals to the chamber cavity of microwave oven

80. What is the use of HV capacitor?

Ans. Used to bypass the high voltage AC to magnetron

81. What is the principle of microwave oven?

Ans. Heating is caused by the vibration of water molecules present in the food

82. Which is not suitable for microwave cooking?

Ans. Metal or metal coated utensils

83. What are the types of microwave oven?

Ans. Mechanically controlled and microprocessor controlled

84. What are the classifications of microwave oven based on their working method?

Ans. Basic model, combination model, convention model

85. What additional parts present in convention model than combination model?

Ans. One fan and a thermostat

86. What are the two types of heat produced in combination model?

Ans. Conventional heat and microwave heat

87. What is the use of auto cook?

Ans. Used to auto cook by applying the weight of food

88. Classify washing machines based on wash methods?

Ans. Pulsator wash, tumbler wash agitator wash

89. Classify washing machines based on convenience of operator?

Ans. Semi automatic, fully automatic, fuzzy logic

90. What are the two types of semi automatic washing machine?

Ans. Single tub model, twin tub model

91. Classify washing machines based on loading of clothes?

Ans. Top loading type and front loading type

92. What is front loading type?

Ans. Clothes are loaded to drum through the opening in front side of the machine

93. Classify the timers present in washing machine?

Ans. Mechanical, electrical, electronic timer

94. What are the timers used in semi automatic washing machine?

Ans. Wash (wash timer), spin (spin timer)

95. What is the use of filter in washing machine?

Ans. This purifies water by collecting dirt and impurities at drain grill

96. What is the use of water inlet?

Ans. Used to connect the main water supply to the wash/ spin tub

97. What are all the modes present in fully automatic washing machine?

Ans. Individual modes, automatic mode, fuzzy mode

98. What is automatic washing machine?

Ans. It can perform washing, spinning and rinsing operations in the same tub

99. What is individual mode?

Ans. It can wash, rinse and spin automatically, can operate fully automatic as semi automatic

100. What is the use of balancing spring in automatic washing machines?

Ans. It reduces the vibration during the wash/spin takes pla